A NEW HOPE

A NEW HOPE:
Wolfhart Pannenberg and the Natural Sciences on Time

By

Stephen Lakkis

CAMBRIDGE SCHOLARS
PUBLISHING

A New Hope:
Wolfhart Pannenberg and the Natural Sciences on Time,
by Stephen Lakkis

This book first published 2014

Cambridge Scholars Publishing

12 Back Chapman Street, Newcastle upon Tyne, NE6 2XX, UK

British Library Cataloguing in Publication Data
A catalogue record for this book is available from the British Library

ISBN (10): 1-4438-6630-X, ISBN (13): 978-1-4438-6630-9

CONTENTS

ACKNOWLEDGEMENTS

Few books are written in isolation. Rather they represent the fruits of many conversations, as well as stages and cycles of feedback and response. This volume in particular developed over many years and grew for the better through the support, insights and encouragements of friends and colleagues. As this book began life as a doctoral thesis, my first thanks go to Prof. Michael Welker at the University of Heidelberg, Germany, who not only supervised the project but did so with a lightness of touch that gave me the freedom to pursue my own insights and to travel my own roads. Prof. Welker's wisdom and guidance, as well as the international and interdisciplinary environment he fostered in Heidelberg were the foundation on which both I and this project could flourish. A deep debt of gratitude is also due to my good friends Markus Höfner, Heike Springhart and, above all, Günter Thomas, for their wisdom and constant encouragement. Not least, my thanks go to my sons for their endless patience, and to my wife for her perseverance.

I am also grateful for the financial assistance that this project received: from Gordon Dicker and the "With Love to the World" Foundation, who enthusiastically supported the early years of this project, and for the generous support of the *Diakonishes Werk* of the *Evangelische Kirche in Deutschland* (in particular Johann Schneider, Sylvia Karlev, and Katharina O'Shea) who overcame considerable hurdles to provide financial support for the closing years of my doctorate.

A special note of thanks goes to Carol Koulikourdi at Cambridge Scholars Publishing, for her infinite patience.

Finally, humble thanks must also go to Prof. Wolfhart Pannenberg, who forever radicalized me for the Gospel of Jesus Christ.

Stephen Lakkis
Taipei
Good Friday, 2014

INTRODUCTION

A NEW HOPE

This book touches on a deep aspect of the Christian life and faith, namely that peculiarly optimistic hope that we carry for the future, a continually new hope in a future that will not only bring about transformation in history and creation, but will also ultimately reveal a final meaningfulness to our current historical experiences of suffering and evil. Of course, this is a hope shared by many people outside the church, who also wait for the future to bring something radically new that will improve the condition of their lives, bring happiness and joy, and end the brokenness in their societies, personal relationships, and own hearts. In contrast to the difficulties and struggles of the present, the future stands as a realm of openness and possibility, and we all wait expectantly in history for that hope to encounter its fulfilment.

The root of our Christian hope in the future is buried in an ancient Jewish witness to the God who acts both unexpectedly and in transformative ways in history. The old stories of the biblical traditions reveal a God who introduces into time contingent and novel experiences of liberation and hope: the promise to an elderly couple of children and a future home; sudden and unexpected liberation and freedom from persecution for those who were enslaved; the promise of a future renewal for a people scattered and exiled. These contingent and unexpected experiences of hope that the community of faith encountered together reached their consummation in the astounding event of the death and resurrection of a beloved Son. In the experience of this extraordinary, sudden, and radically new event of the resurrection of Christ, everything was transformed. History and its significance were changed, and through this startling, novel event life was infused with a sudden and unanticipated sense of hope.

Yet since the Enlightenment, the secular hope in the advancements of history as well as our Christian hope in the radically contingent, salvific, transformative and liberating action of God in time, were undermined and even negated by the rise of a new scientific view. This new understanding of time and the universe rejected temporal and historical contingency in favour of a world view based on the absolute lawfulness of natural processes, a lawfulness which discarded the idea of radical novelty in history, and turned its back on unexpected divine action in time—especially for any event as radical and "unlawful" as a resurrection from the dead. This ideology of strict lawfulness and the associated ontological absence of contingency found deep support in the amazing predictive power of classical Newtonian physics, and reached its pinnacle in the twentieth-century with relativity theory's understanding of the universe as a determinate, and unchanging four-dimensional solid. Suddenly, we were faced with a curious ideological tension. On the one hand were the historical and theological sciences (together with popular opinion) which still worked with a belief that the future was open and free at least in some sense to be written or changed. On the other hand, the physical sciences were operating with an understanding of the strict lawfulness of the universe, of the power of science to calculate that future using natural laws, which determined how the past shaped the present and in turn would concretely form the future. The fused, four-dimensional space-time of relativity theory then set about describing the universe as an invariant block which excluded radical contingency, sudden novelty, and future change. Society was then left with a deep tension, if not open conflict, in the fundamental ways that it perceived time and history. If the scientific models were correct, did we still have good reasons to expect novel events from the future? Could we really continue to see the future as an open and unwritten stage for unique and unprecedented human and divine action? In the context of such belief in a strictly lawful, determinate universe, was there any room left for a Christian hope in indeterminate, radically contingent and transformative divine actions and events in time? Was there even room for new hope in history?

In this book I want to analyse this point of friction by examining the work of German theologian Wolfhart Pannenberg and his understanding of time. What makes Pannenberg's work so interesting in this context is not only his impressive scholastic rigour and his novel approach to a theological understanding of time and history, but his attempt to provide a theological theory with broad objective, universal validity. In the interdisciplinary discussion with the natural sciences, Pannenberg takes his skills as a theologian and attempts to construct an understanding of time which not only incorporates theological truths but which would also account for scientific aspects of lawfulness and regularity. Whereas other theologians may be content to limit their studies and reflections only to the field of theology, Pannenberg pursues a long-term interdisciplinary project that aims to offer results that are universally sufficient both for theology and the natural sciences. In his study of time, the crux of the issue is whether in the end we are finally subjected to the all-determining laws of nature or are reasonably able to place our hope for the future in the work, guidance and activity of God.

Over the following chapters we will discover that Pannenberg understands time as a direct expression of God's absolute and omnipotent power and rule over creation. This absolute power of God calls into question scientistic claims about the lawfulness of the universe. In presenting such an argument, Pannenberg's position is genetically connected with a medieval doctrinal position based on the absolute power of God (*potentia Dei absoluta*). Furthermore, we will find that since Pannenberg shares an understanding of time as both linear, objective, and determined, his work on time is not diametrically opposed to the concept of time offered by theoretical physics. However, whereas classical physics stresses the rule of natural laws in a system of time which extends from past to future, Pannenberg stresses the ultimate and absolute rule of God in a system of time which moves from future to past. God's faithful and absolute rule finds its location and expression in God's continuing creation of every discrete moment of time and history.

In this respect, Pannenberg presents a novel approach to the interdisciplinary discussion of time. But unfortunately it is not without serious problems. We will see that Pannenberg fails to adequately understand or sufficiently critique the current scientific models of time. More importantly, Pannenberg's stress on God's absolute power and rule over time launches an arms race between theology and the sciences, with the spoils being absolute control over the physical universe. Furthermore, Pannenberg's approach serves to instrumentalize time, transforming it into a weapon of divine punishment. This leaves him unable to respond authentically to experiences of suffering, sin and evil in time (a problem the sciences avoid since they do not hypothesize the existence of a good God) or to deal adequately with time's own state of fallenness. We will see that the bulk of these problems stem from Pannenberg's insufficiently nuanced understanding of God, whom he sees starkly and necessarily as an "all-determining reality". I will argue that Pannenberg's approach, with its focus on the absolute power of God, resurrects a concept of God developed by Thomas Bradwardine in the fourteenth century and used in the conflict against Averroism's own conception of natural, physical determinism.

I will also argue that both Pannenberg's theory of time and the deterministic scientific view of the universe share a common weak point, namely a reductionistic understanding of the concept of lawfulness. Therefore, this volume will close with two main arguments: (1) that theology must consistently incorporate the insights of a *theologia crucis* into its work on time if it wishes to engage satisfactorily with the fallenness of time, and avoid skewed ideas of determinism based on God's absolute divine power and rule. (2) Both theology and the sciences could benefit from a more differentiated understanding of the concept of lawfulness—an argument I make by turning to the distinction in jurisprudence between "rule by law" and "rule of law". This conceptual reworking of the concept of lawfulness is already well underway in quantum physics; our theologies would also benefit from adopting this distinction.

The main story of this volume—the critical analysis of Wolfhart Pannenberg's work on time in discussion with the natural sciences—progresses in six parts. It will be helpful here to offer a brief overview of the major plot points along the way.

1 Discussing Time in a Postcolonial Context

From the perspective of the natural sciences, an examination of the concept of time may strike us as a purely theoretical one, rooted in scientific objectivity and described by the rigorous language of abstract mathematics. But outside the natural sciences, this view has attracted critique. The ways in which we understand the concepts of time and history reflect a deep cultural engagement with the "problem of time" itself. Various cultures in various ages have had widely different understandings of time and the way that past, present and future interrelate. Even something as simple as the mental diagram which pictures time flying like an arrow from left to right, with the past behind us and future in front of us, falls apart in various cultural contexts (such as in Taiwan and China) where we stand with our backs towards the future.[1] While it is therefore a given that the majority of Western theological and philosophical discussions of time will occur from a Western viewpoint and thus be coloured by Western assumptions about reality, it will be helpful for us to thematize the issue of culture and cultural perspectives at the very beginning. Chapter One sketches out some of these difficulties facing an investigation of time in a postmodern and postcolonial context.

This examination is particularly important given that Pannenberg's stated theological aim is to provide a theology with "universal validity",[2] in the same way that the sciences are seen to offer objective "universal validity" in their models. The foundations of Pannenberg's own work on time (laid in the 1950s) predate an awareness of postcolonial issues. As an example of the problems this can raise, we will see that Pannenberg remains heavily dependent upon a modernist historiographical concept of neutral or objective "universal history"—a position which, though shared in part by the natural sciences, has now become academically awkward. Western attempts to portray history as a single, unified story have become linked with the suspicion of cultural and political dominance.[3] Therefore the question whether such an "objective" or neutral perspective on time exists (either in theology, philosophy or the natural sciences) will need to be examined briefly.

---

1 In Mandarin, when discussing spatial locations, *qián* (前) signifies "in front of" while *hòu* (後) signifies "behind". When discussing time, *qián* signifies the past (*qiántiān:* the day before yesterday) and *hòu* the future (*hòutiān:* the day after tomorrow).

2 Cf. e.g. Wolfhart Pannenberg, *Anthropology in Theological Perspective,* trans. M. J. O'Connell (Edinburgh: T & T Clark, 1999), 15.

3 Pannenberg's work on universal history was heavily influenced by the important role given to such a unified history in Gerhard von Rad's theology of the Old Testament. Modern biblical scholarship is revealing the extent to which von Rad's work practises just such a reductionism, forcing the Old Testament sources into a universal historical model and excluding sources which are critical of that universalist approach. Pannenberg's adoption of von Rad's universalist approach is understandable given the era in which he was working, but it makes the reception of his ideas more difficult for us today.

One of the aspects that makes Pannenberg's approach interesting is his claim to universal validity, and his conviction that theology should not only strive for internal coherency but should also seek agreement with the external descriptions of reality provided by other disciplines.[4] To his credit, Pannenberg is not interested in withdrawing theology back into a "safe harbour" disconnected from the truth claims of the sciences, and he is critical of those theologians whom he feels have taken this approach. If we take Pannenberg's claims to universal validity and universal coherence seriously, then this external coherence offers us a vital criterion for assessing Pannenberg's work on time. Is Pannenberg's work able to satisfactorily account for the phenomena and descriptions of time witnessed to by other disciplines? And if we draw here on Ockham's Razor, then to what degree is Pannenberg's theological concept *better* able to account for these phenomena, and thus supersede those other theories rather than just duplicating them?

We may suspect that theological conceptions of time, with their division of time into "earthly temporality" and "divine eternity", purposefully create such a safe harbour, keeping the sciences out of theology's private discourse on the eternal. But Pannenberg rejects this type of division and argues instead for a unification of all time into a single "omnitemporality". As a result, one the one hand scientists and theologians can share the same object of enquiry and participate in the same discourse. However, on the other hand an overly quick unification of these concepts begins to muddy the long-standing historiographical distinction between *Historie* and *Geschichte,* between the underlying ontological structure of time and its content. The tacit agreement between classical physics and the philosophy of time is that their own fields attempt to deal with the objective structures (ontology) of temporality. This then relegates the historical or social sciences to a second order examination of subjective occurrences in time. The development of relativity theory has overcome this distinction by fusing the ontic structure of space and time together with their contents. Yet this has reinforced the conviction of the sciences that they can once again begin writing an overarching metanarrative for an assumed objective, universal, "cosmic history"—at the precise mo-

4 Wolfhart Pannenberg, *Systematic Theology,* trans. G. W. Bromiley, 3 vols. (Grand Rapids, MI: Eerdmans, 1991–97), ii. 21–2.

ment when many postmodern and postcolonial voices seek to free themselves from the dominance of such a single metanarrative.

Today we are highly aware of the role that culture and biography play in our academic discussions. We need to bring the same awareness to our discussion of time. Indeed, early in Pannenberg's career James Robinson insightfully noted that Pannenberg's personal experiences of the Second World War were influencing the development of his theology.[5] This insight prompts us to introduce a very brief biographical sketch of Pannenberg's early life, noting his strong focus on medieval theology and the concepts of God's "power and judgement" over the world, as well as his disillusionment with the "liberal democratic tradition".[6] Robinson's insight into the importance of Pannenberg's early biography will be an important key in understanding Pannenberg's work in this study. While much of the current literature examining Pannenberg's work focuses narrowly on his *Systematic Theology*, in this volume I will argue that it is particularly in Pannenberg's earlier works and essays where we find the formative material that grounds, shapes and gives vital context to his work in the *Systematic Theology* and beyond. If we want to appreciate the impressive consistency of Pannenberg's thought and the full arc of his thinking on time, then we must examine the roots of his thought, rather than simply performing a truncated analysis of his *Systematic Theology* alone.

2 Categorizing Temporal Systems

The intercultural and interdisciplinary discourse on time must deal with a wealth of differing temporal systems. Thus Chapter Two will provide an overview of these different approaches, and offer a simple guide for grouping and categorizing these systems so that we can better map and locate Pannenberg's work on the broad horizon of discourse on time. The

---

5 James M. Robinson, "Revelation as Word and as History", in id. and John B. Cobb, Jr. (eds.), *Theology as History* (New Frontiers in Theology: Discussions among Continental and American Theologians, 3; New York: Harper & Row, 1967), 1–100 at 3.

6 Wolfhart Pannenberg, "God's Presence in History: How My Mind Has Changed", in *Christian Century*, 11 March 1981, 260–3 at 261, 263.

method I will argue for is based on the vectors of linearity, objectivity and reference. *Linearity* refers to the vectorial nature of time: the degree to which it represents a linear extension in one direction. The opposite pole of a linear temporal system is represented by circular temporal systems (typical in many Eastern models of time). The second measure, *objectivity,* refers to the hypothesized "location" of time: i.e. whether time is a construct within human consciousness (*subjective*) or whether it has its own ontic basis outside of human consciousness (*objective*). The final measure is *reference.* Simply put, whereas objectivity measures the location of time, reference measures its "number". If a model views time from a singular, unified and universal frame of reference, we can describe it as unireferential; if the perception of time is inherently divided across multiple, equally valid frames of reference, we can class it as multireferential. While these three measures are necessarily reductionistic, together they provide a helpful basis for categorizing a range of temporal systems, and also help to highlight points of similarity and difference between systems. Using this categorization, we will be able to see that whereas special relativity presents time as linear, objective and multireferential, we can classify Pannenberg's concept of time as linear, objective and unireferential.

The danger behind developing a guide for categorizing and assessing varying concepts of time is that it assumes that we actually have a range of concepts to work with. In reality, this is becoming less and less the case. The dominance of scientific views on time has left little significant space for competing conceptions. What this means *de facto* (to the disappointment of philosophers, theologians and historians) is that in popular and academic discussions, science stands as the final authority and arbiter for our understandings of the reality of time. In this context, can theology even re-enter the discussion of time or create a new space in the discourse where alternate theories can be advanced? The common tendency to segregate the sciences and theology into mutually exclusive academic discourses (for example, under the assumption that the sciences speak the language of "fact" while theology speaks only the language of "value"[7]) limits theology at best to discussion of some

---

7 Cf. Ted Peters, "Science and Theology: Toward Consonance", in id. (ed.), *Science and Theology: The New Consonance* (Boulder, CO: Westview, 1998), 11–39, esp. 17–18; Ted Peters, "Naturwissenschaft und Religion. Ein wachsender

secondary form of "theological time". At worst, it forcefully silences and removes theology from the discourse. Yet, again, if we are willing to take seriously Pannenberg's own demands regarding the possible universal validity of theology and its truth claims, then such a segregation need not be accepted. Schools of theoretical physics admittedly dominate the discourse on time, yet theology has a special role in creating a space in the discussion for highlighting what it sees as the shortcomings of those scientific concepts. Without this initial step (which forms the bulk of Chapter Three) theology would remain unable even to voice the relevance of its own contributions to the discussion, or raise reasons why it should be readmitted to the discourse.

By the same token, despite the general dominance in the West of objective theories of time, we should not completely shut out the important counter-arguments raised by the proponents of subjective theories of time. Their insights offer valuable opportunities to temper our views of objective time. Therefore, Chapter Two closes with a brief overview of the insights developed in key subjective theories of time. On the one hand, this provides better contrast for the following discussion of objective time; on the other hand, it will make clear why those subjective theories have failed to gain significant traction in the Western discourse.

3 Making Space for Time

In a context dominated by scientific conceptions of time, is there really any way for a theologian to regain access to the discussion? If we are unsatisfied with a conceptual segregation that relegates theologians and scientists into different corners of the room and demands that each group play with their own form of time, is there then any way that theologians can make space in the scientifically dominated discourse and reintroduce a theological perspective? What is needed here is a careful analysis of the insufficiencies of the scientific models, together with ways that a theological conception could address those problems. To do this, we need to engage with and critique scientific models of time within their own contexts. Chapter Three aims at just such an engagement, offering

Forschungsbereich", *Verkündigung und Forschung*, 49 (2/2004): 52–72, esp. 57–8.

a brief philosophical (rather than detailed mathematical) overview of Minkowski space-time and Pannenberg's own critique.

This type of philosophical engagement between physicists and theologians already has a long and valuable history.[8] Using this method from the history and philosophy of science, Chapter Three will draw upon texts from the watershed period of development in relativity theory (1904–1927). We will examine the impetus for the development of space-time theory, namely difficulties which began to arise with the concept of stable frames of reference within classical physics. We will find that rather than continue with a Newtonian conception where time was absolute and formed the container within which the physical universe "moved",[9] Poincaré and Einstein disassembled belief in the absoluteness of physical measures, including time, and replaced such absolutes with a relativistic system. It was then Hermann Minkowski who discovered the means to regain the "absolute" nature of the universe by re-combining spatial and temporal dimensions into a single invariant solid. In this way he provided the basis for the manifold theory of space-time and the "block universe".[10]

The development of Minkowski space-time had a far-reaching impact on the philosophical discussion of time in physics, philosophy, and theology, particularly with regard to the philosophical issues of restriction, invariance, and determinateness in the universe. The inherent restrictions of space-time redefined the future away from being a field of open and infinite possibility and limited it to a description of an object's already existent extension in four dimensions. Space-time's stress on the integrated nature of temporality and spatiality also effectively limited a theology which tried to understand God as active in time yet lacking spa-

8 It was clear from the beginning that new physical theories on the nature of time would immediately impact on theological and philosophical understandings of reality. We see this e.g. in the impressive early work of Arthur Eddington, *Space Time and Gravitation: An Outline of the General Relativity Theory* (Cambridge: CUP, 1920), and especially in his Gifford Lectures, Arthur Eddington, *The Nature of The Physical World* (1928; London: Dent, 1964).

9 Isaac Newton, *The Mathematical Principles of Natural Philosophy* (3rd edn., London: Motte, 1729), 9–10.

10 Hermann Minkowski, *Raum und Zeit: Vortrag, gehalten auf der 80. Versammlung Deutscher Naturforscher und Ärzte zu Cöln am 21. September 1908* (Leipzig: Teubner, 1909).

tial corporeality, and also raised problems for a theology of divine omnipresence (which assumes presence *across* various frames of reference rather than *within* a single one). Invariance stresses the static nature of space-time, or the inability of time to flow (since it lacks another temporal metadimension in which it can move). Philosophically, this leads to the radical permanence of matter in the Minkowski universe. The invariance of this extension in four dimensions also suggests that the "future" state of these objects is already set and unchanging, which brings us to the problem of determinateness. Since the "future" already exists within the space-time manifold, we seem to be left with a rejection of any (human or divine) interventionist activity in history.

It is here that Pannenberg launches his critique of the space-time model, providing two points of critique. First, Pannenberg opposes scientific reductionism which focuses on small parts of a bigger picture. He reiterates his long-standing argument that parts necessarily presuppose a whole, and that by concentrating on the divisions of time, physics has overlooked the necessary pre-existing unity of all time. Unfortunately, we will discover the limitation of this critique, arising from Pannenberg's misunderstanding of the absolute and invariant nature of the Minkowski universe. Pannenberg's second critique stresses that absolute space-time should not be correlated with God's eternity, and that the space of relativity is not "the space of God's omnipresence". While the theological concern here is to avoid the dangers of pantheism, we will see that it is unclear whether Pannenberg is presenting an ontological or an epistemological critique of relativity. The ontological option presupposes a division between a divine space and time and the absolute space-time of relativity, avoiding pantheism but also losing the concept of God's omnipresence. Furthermore, by appealing to different forms of space and time, Pannenberg reintroduces an ontological distinction, leading us back to segregated concepts of theological and scientific times. The epistemological option views space-time simply as a geometrical concept, one unable to accommodate God's reality. Here again, if the epistemological systems of theoretical physics are unable to understand time as theology does, then we once again have a failure in the basic premise of the interdisciplinary discussion: that we can both investigate the same object and arrive at some sense of universal validity.

This is not to say that Pannenberg is wrong in his intentions to critique the dominance of relativity theory, merely that his attempts fail to hit the mark or stay true to his own theological motives. I will, however, attempt to extend and strengthen Pannenberg's critique by highlighting limitations in the model, specifically by pointing to non-theological aspects of reality that relativistic mechanics is unable to incorporate. I will argue that as a system of mechanics, relativity is useful for understanding the extension of substantial realities, but is finally ill-suited for understanding non-substantial realities such as information and pattern (drawing on Whitehead), language (J. L. Austin), as well as pain and consciousness. Rather than take issue with Minkowski space-time itself, I will argue instead against those philosophers and theologians who misapply the model, promoting a reductionistic physicalism that sees space-time as a sufficient system for describing the "totality of all that is".

But there are not only points of critique to be made between physics and theology. Curiously, through space-time's unification of all time into a single universal history, which is then combined with all reality to form a single unified whole; in its view of the universe as bound within an (eternally) determined course; in its rejection of efficacious change within that unified "reality–temporality" (so much so that universal change can only occur via the destruction of one block universe and the creation of another), we find that the four-dimensional, invariant, block universe of Minkowski space-time actually carries all the hallmarks of a biblical apocalyptic world view.

4 Laws of Nature, Rule of God

In Chapter Four, we will find that Pannenberg's concern with the dominance of the natural and historical sciences represents a long-running motif in his theology. He is particularly concerned that their epistemologies and academic approaches can lead to an unjustified rejection of the doctrines of the Christian faith. One aspect of the physical sciences that deserves detailed attention, he believes, is the Newtonian concept of inertia. Since inertia provides a determined, lawful and mathematical basis for tracking the progression and location of an object into the future, and provides objects with their own continuity across time, Pannenberg sees the theory of inertia as impacting upon doctrines of the direct divine control of the natural world.[11] Thus inertia effectively offers a form of creaturely independence over against God which not only impinges on God's absolute control but also constricts the future, making it regulated and predictable. The rule of natural laws removes contingency from the future and makes the future a wholly knowable state based on the present. Such a regulated future then allows no space for divine contingencies, particularly a resurrection from the dead. Thus for Pannenberg, dealing with the problems raised by the principle of inertia stands as one of the most crucial tasks to be addressed in the dialogue between science and theology.[12] If we accept the rule of natural laws, then time becomes a mere development, the growth into the future of a seed that was already planted in the past. Such an entelechy would not allow for true contingency, and the future would never contain anything qualitatively new.[13]

But Pannenberg's critique extends beyond the natural sciences. He raises particular concern about the way the ideology of strict lawfulness has infiltrated the historical sciences, centred around the principle of analogy which argues that the historian can only accept as true those historical events which correlate to his or her own experiences (or the accepted experiences of others). Pannenberg fears that if analogy becomes the criterion by which the truth of historical events is determined, then there is no chance to accept anything truly novel occurring in history. It would also be impossible on these grounds for Christians to defend the historical truth of a radically novel event such as the resurrection.[14] Thus Pannenberg's aim is to defend theology against scientific and historical models which stress a universalizing law or are based on the absolute rule of universal natural laws. These models destroy an understanding of the contingency of the universe and thus either lead people to reject

11 Wolfhart Pannenberg, "Theological Questions to Scientists", in id., *Toward a Theology of Nature: Essays on Science and Faith*, ed. Ted Peters (Louisville, KY: Westminster John Knox, 1993), 15–28 at 20.

12 Pannenberg, "Theological Questions to Scientists", 20.

13 While it has become common to read Pannenberg as a Hegelian, we will see that Pannenberg's strenuous critique of any entelechy within history places him in opposition to Hegel.

14 Cf. Pannenberg, *Systematic Theology*, ii. 285.

doctrines of God's contingent action in creation, or promote a form of deism which prohibits God from acting in time now.

In response, Pannenberg enters an arms race against these deterministic natural laws, seeking to trump and overpower them with an even greater, more powerful Ruler: the absolute power of the all-determining God. Thus the development of Pannenberg's concept of time will revolve around a defence of this doctrine of God.

For Pannenberg, if, via inert and the rule of natural laws, the past determines the present which then determines the future, then scientific concepts of inertia and natural law infringe on the power of God and must be rejected. If the development of the universe is determined by natural laws then theology would no longer be free to speak of the contingent acts and decisions of God in history. In response, Pannenberg offers the following thesis: It cannot be the past which determines the present and future via the rule of natural laws, but rather it is the future which must determine the present and the past, based on the rule and faithfulness of God.[15] In a reversal of our normal understandings of time, Pannenberg argues that God creates each new discrete moment through the absolute power of his divine will and control, and then places those new moments into time to form the present. In this process of backward linkage (*Rückbindung*), each new present moment is a direct act of God's continuing creation of time which God then links back to each immediately previous present, ensuring the coherency of the new moment of time with its immediate past. Thus for Pannenberg, what we call the past is simply the remnant of previous free and contingent acts of God being pushed from the future into the past. Since the direction of time comes from the future, the residue of time which forms the past lacks any determinative power over either the present or the future.

Five points are vital for understanding Pannenberg's concept of time. First, time is linear, but sterile. Time progresses linearly from future to past, but because the progression of time is reversed no temporal event can form the basis of any subsequent future moment. If time were not sterile, it would lead to an entelechy. Second, the concept relies upon an imperial model of the absolute sovereignty and rule of God to produce

15 Wolfhart Pannenberg, "Theology and the Kingdom of God", in id., *Theology and the Kingdom of God* (Philadelphia: Westminster, 1977), 51–71 at 59.

each and every temporal event. For Pannenberg, this absolute, dominating rule of God over time and the universe is non-negotiable, and it is the marker of his doctrine of God: "The deity of God is his rule".[16] Third, the truth of time is not located in human consciousness, but in the powerful acts of God. Thus time is objective. Fourth, since time stems from God's absolute and omnipotent will, *all* temporal events (both natural and historical) are radically contingent at an ontological level and come directly from the hand of God. Thus, fifth, if the historical or natural sciences experience a thread of continuity and coherency between two events, such continuity must be based solely upon God's will and desire for that continuity to exist. Continuity in nature and history is thus disconnected from any sense of natural lawfulness. God's faithfulness and power replaces inertia and natural law as the basis and driving force of historical continuity. Conversely, if historical continuity is *not* maintained, Pannenberg interprets the resulting destruction as God's judgement.

A doctrine of God infused with the concepts of divine power and control forms the foundation of Pannenberg's understanding of time, and the present in particular. For Pannenberg, the present tense is defined and delimited by experiences of control, and since he sees all creation as functionally open to God's control, God thus perceives time as an "eternal present". Eternity (the gathering together of all time under God's power and control) thus becomes the ontic "truth of time", and (using the language of special relativity) God's divine frame of reference.

Thus by stressing (a) the absolute rule and sovereignty of God, (b) God's absolute control over time and all temporal events (either in creative support or in destructive judgement), (c) the lack of any determination of the past over future events, and (d) the final impotence of what the sciences claim to be determinative natural laws, Pannenberg replaces a scientific "rule by natural laws" with a theological "rule by divine power", swapping out a form of temporality based on natural determinism with one based on divine determinism. In this way, Pannenberg can defend the activity of God in time and creation by promoting an image of God as an ultimate and unchallengeable power, outbidding the laws of nature with the *potentia Dei absoluta,* the absolute power of God. But is

16 Ibid. 55.

such a move defensible? Is it not theologically flawed and Christologically untenable?

5 The Ultimate Cause of Evil

In Chapter Five I will argue that Pannenberg's work on time is in essence a reworking of the thirteenth- and fourteenth-century debate with Averroism, particularly the appeal to the *potentia Dei absoluta* in defence against Averroism's teachings on cosmic determination. This approach was championed especially by Thomas Bradwardine, an approach we can follow through Pannenberg's own analyses of Bradwardine's thought. Pannenberg, as a scholar of medieval theology, explains how Averroism presented Christian theology with an influential, academic system which argued for a process of natural cosmic determinism. Because it effectively limited the immediate activity of God in the natural world, it was perceived as a threat to doctrines of the sovereignty of God.[17] At the same time, early Scholastic realism was engaged in the pursuit of "universal and eternal truths", which also effectively endangering the concept of "historical particularity". The response among Nominalists and the Oxford-based *schola Augustinia moderna* was to shift focus from historical ideals to historical particulars, and to appeal to God's *potentia absoluta* as the basis both for all such contingent, historical particulars[18] as well as for their integrated continuity in the natural world. But it was Thomas Bradwardine who pushed this dependence upon the *potentia Dei absoluta* to its final end, extending its jurisdiction as truly *absolute* to all temporal events: to all events in the lives of human beings, and to

---

17 Wolfhart Pannenberg, review of "Gordon Leff: Bradwardine and the Pelagians. A Study of his 'De Causa Dei' and its Opponents (= Cambridge Studies in Medieval Life and Thought N. S. vol. 5). Cambridge (University Press) 1957. 282 S.", *Zeitschrift für Kirchengeschichte,* 69 (1958): 355–61 at 359, 357; hereafter as "Bradwardine and the Pelagians".

18 Wolfhart Pannenberg, review of "Heiko Augustinus Oberman: Archbishop Thomas Bradwardine. A Fourteenth Century Augustinian. A Study of his Theology in its Historical Context (= Diss. Utrecht 1957). Utrecht (Kemink en Zoon) 1957. XI, 246 S., hfl 10.–.", *Zeitschrift für Kirchengeschichte,* 72 (1961): 173–5 at 174; hereafter as "Archbishop Thomas Bradwardine".

all temporal phases, past, present, and future. In so doing, he defended God's sovereignty and power as well as the historical particularity and contingency of all events.

We find here an astonishing number of parallels to Pannenberg's approach in his response to the natural sciences. Pannenberg enters into a discussion where he sees modern theologians challenged by the dominant physical sciences which promote a natural form of lawful determinism (either via the principle of inertia or the invariable space-time manifold), and exclude the immediate activity of God in time. Furthermore, the focus on universal truths (in the form of natural laws or historiography's analogical method) led to a scientific rejection of novel and contingent events, and a historiographical rejection of particularity. Thus I will argue that we see in Pannenberg's response to the natural sciences is a return to Bradwardine's theology (via Martin Luther), characterized by an appeal to the absolute power and rule of God and a divine determinism, all in order to protect faith against an extra-theological natural determinism. His work on time revives a stress on the absolute power of God and the inability of history and nature to provide coherent development in time on their own. It argues for God's power over nature, God's creative ability to call into existence and maintain naturally disconnected, radically contingent, and historically particular events. It asserts God's power to link these events into a coherent order in accordance with God's faithfulness. It draws heavily on the absolute power of a God who expresses absolute control over all time from the realm of his omnipotent, eternal present. And his approach proclaims the absolute sovereignty of a God whose power defines his very being. In essence, Pannenberg resurrects the Bradwardian method and its unyielding stress on the divine *potentia absoluta*. Yet we will see that this leaves Pannenberg vulnerable to the same problems that Bradwardine encountered, most notably the final attribution of all sin and evil to God.

In his early work, Pannenberg indeed accepts that God is the final cause of sin and evil.[19] Following Luther and Bradwardine, he also argues that the concept of human free will (*liberum arbitrium*) was used in error by the church fathers to stress a human role in sin. Such a position clash-

19 Wolfhart Pannenberg, "Christliche Glaube und menschliche Freiheit", *Kerygma und Dogma*, 4 (1958): 251–80 at 275.

es with the claim that God is the sole subject and actor of history. To flesh out this idea, Pannenberg turns in his later work to social and cognitive psychology to find a method for attributing sin to the human being without the need to include historical or temporal action. He focuses on the concept of egocentricity, which he can then use to distance the concept of personhood from its embodied dimensions and from the physical actions that such an embodied individual may perform. This disconnects the concepts of sin and evil from expression in physical events and rather locates them solely in cognitive-psychological states.

I will argue that Pannenberg's attempt fails for several reasons. First, his psychological reinterpretation of sin as a pietistic, personalized, individualistic egocentricity fails to do justice to the biblical testimony of the broad dimensions of sin and evil, or even to our experiences of corporate and cultural evils which transcend individuals and tyrannize and oppress them. Second, it abandons a nuanced concept of the human being in its full corporeality, and confuses a cultural context for a universal, genetic trait. Third, and most seriously, Pannenberg's reworking of time and causality disrupts our basic ethical understandings of action, and this disruption spills over into the field of jurisprudence. In a legal setting, if there is no historical coherence, then it is meaningless and even perverse to speak of an ethical matrix of causality and responsibility which lead to legal attributions of blame and subsequent judgement. The entire range of historical evils and injustices that the law seeks to constrain must finally be revealed as instances immediately created by the all-determining hand of God.

In Pannenberg's neo-Bradwardianism, we find that he does indeed reinterpret all temporal moments of death and suffering as expressions of the immediate will and action of God. Thus he is led to the same problems Bradwardine encountered. First, since suffering comes from the hand of God, the reality of such life-destroying evil is relativized or denied by claiming that such experienced suffering must ultimately be good if only perceived from God's perspective, if viewed with eternity as the truth of time. Second, such a reworking serves to instrumentalize time, specifically the arriving, unknown future. The future is transformed into a weapon of God's judgement, bringing heavenly salvation or divine destruction and damnation with each newly created moment. As a result, Pannenberg can no longer see time itself as a creature in need of fulfilment and transformation under the healing power of God. Time is excluded from the need for redemption and becomes simply a tool of divine revelation and judgement.

Finally, in his *Systematic Theology* and later works, Pannenberg attempts to rework the structure of his approach using T. F. Torrance's work on fields of force. Yet while the metaphor of his model has changed, Pannenberg's underlying goals and premises remain consistent. We find the same stress and reliance on God's absolute power and rule, on time that flows from future to past, on a rejection of creaturely independence. God's *potentia absoluta* remains the core around which all else revolves. Therefore, even with the change of metaphor Pannenberg is unable to escape from the range of difficulties that necessarily accompany his idea of God as all-powerful, all-determining reality.

For Pannenberg, God's divine being is always to be correlated with God's ultimate, supreme and absolute rule. The "supreme being is called God in virtue of his all-embracing power and lordship (*potestatem et imperium*). Without these he would not be God".[20] Yet how can such an autocratic and imperialist concept of God cohere with the figure of a Christ who suffers, is tortured and executed? In Pannenberg's defence of God before the powers of the natural sciences, he focuses on God's insurmountable and absolute might, yet in doing so he leaves little room for the development of a satisfactory *theologia crucis.* Any serious theological discussion on the concept of time must take as its starting point not a speculative and philosophical conception of God but rather the One who was incarnated into time and savagely pushed back out of it. If we begin with a *theologia crucis,* if we begin with the perfect self-revelation of God within the framework of the destruction and suffering of life in time, and if we want and maintain a truly Christian hope for redemption and salvation from that suffering in both the present and future, then we must allow for a concept of time which reveals the cruciform nature of all creation, including created time itself. In Christ, God suffers under the fallenness and brokenness of time, suffers under its true power for violent destruction; and through Christ's resurrection God points to a process of healing and transformation of all creation, including time. It is not in the *potentia absoluta* but rather in the weakness of the crucified

20 Pannenberg, *Systematic Theology,* i. 417 n. 173.

Christ that power is made perfect. By pursuing a neo-Bradwardianism that utterly subjugates everything to God (*"Gott gänzlich unterworfen"*), Pannenberg finally fails to escape the traps of divine determinism, and is unable to deal satisfactorily with the unavoidable problem of historical suffering.

6 Rule by Law, Rule of Law

In Chapter Six, this volume concludes by re-approaching the interdisciplinary discussion on time from a new perspective on the concept of lawfulness. Pannenberg's argument with the natural sciences is that they have developed an interpretative framework based on the priority of natural laws and the ability of these laws to specify the development of all temporal/historical events. Pannenberg's own response to this situation was simply to outbid the scientific view of "rule by natural laws" by appealing to a theological "rule by divine decree" based on God's unsurpassable and illimitable *potentia absoluta.* At this point, we can see that Pannenberg's debate with the natural sciences is actually fuelled by a shared understanding of the concepts of rule and law, and who has the rights to that system. But both models are ultimately reductionistic and unhelpful.

In an attempt to sketch a possible way forward, I will argue that we should return to a jurisprudential distinction between "rule *by* law" and "rule *of* law". In broad terms, authoritarian "rule *by* law" refers to those instances where laws are used as an instrument of (societal) determination and control. In contrast to this type of instrumentalization of the law (which seeks to use "lawfulness" as a tool of authoritarian control), "rule *of* law" systems establish a *constrained yet supportive* framework within which free and contingent action can still occur. Under the rule of law, the law is pre-eminent and can serve as a check against the abuse of power, whereas under rule by law, the law serves as a tool of suppression.[21] Classical and relativistic physics as well as Pannenberg himself all advocate a rule *by* law system. They offer an instrumentalization of

21 Cf. Brian Z. Tamanaha, "The Rule of Law for Everyone?", *Current Legal Problems,* 55 (2002): 97–122 at 101.

the law, expressed in the matrix of time. In contrast, a rule of law system would aim at establishing a lawful (constrained), yet supportive framework that still allows for free and independent action: a supportive space for free and responsible action constrained within legal bounds. I close this volume by arguing that we already see these moves to "rule of law" systems in the sciences and theology. Quantum theory displays a growing awareness of the ontological complementarity of constraint and contingency/uncertainty, and modern theologies that emphasize a *theologia crucis* display an understanding of God who submits to the constraints of a fallen temporality, not, in an authoritarian way, to rule *by* it but to transform and redeem it from within. Such anti-dictatorial "rule of law" systems are more difficult to instrumentalize since the sovereign submits to and exists within that system, rather than wielding a system that applies forcefully to all except the sovereign. In the concreteness of a *theologia crucis,* through the suffering and resurrection of Christ in time, God rejects those models which would see him as a distanced, autocratic, absolute dictator. In the suffering and cross of Jesus, we witness God's subjection to time; and in the resurrection of Christ we see the transformation of the fallenness of time, and the hope for a final healing over all time's brokenness.

CHAPTER ONE

EXAMINING TIME IN A POSTCOLONIAL CONTEXT

1.1 Neutrality and Universality

It has become common practice to begin these types of academic studies with a brief definition of key terms, followed by a list of the various scholars who have associated themselves with each view, an outline of those views, and then a simple declaration of the particular definitions which will be used in the following discussion. Such a clichéd beginning is not an option for us. The challenge we have ahead of us here is to understand and define both the concept and experience of time, particularly as it is seen in the work of the German theologian Wolfhart Pannenberg, and in the natural (particularly physical) sciences. Thus a definition of time—one that does justice to our experiences of it—is our goal rather than our starting point. Nevertheless, a few guiding comments will help us to navigate the discussion, and keep us from getting lost in a debate that has spanned contrasting cultures, various academic disciplines, and several millennia.

In this chapter, I simply want to sketch out some preliminary notes on aspects which influence the way we form a concept of time, such as our societal and scientific presuppositions, the influences of culture, and our personal biographies. This is a vital step. If we fail to appreciate these influences, we will tend to assume that concepts of time simply develop in a neutral and objective way, and (worse) that all concepts of time are essentially universal: that the natural and correct view of time is just how physics describes it and how Westerners interpret it. But any investigation of time which takes such a "normativist" approach will soon run into a number of problems; it will have difficulties accounting for the deep cultural differences evident in various theories of time (and the concepts of history which grow out of them), and will wonder why so many people are so "naively" reluctant to adopt these proper, normative, scientific views of time (and the imperialistic understandings of time and historiography they often promote). This is not to say that the search for universal normativity is itself inherently destructive or undesirable, simply that we need to beware of its abuses. In the same way, completely rejecting any sense of normativity or universal aspects of truth will simply lead us into a state of fragmentation where none can communicate meaningfully with those outside their own group. In an interdisciplinary discussion, we need to tread finely between these poles of subjectivity and claimed objectivity, between untranslatable fragmentation and the search for unified truth about the world we share together. This stress on the shared nature of universal truth is the starting point for Pannenberg's theology, and the framework which drives him to pursue interdisciplinary discussions with the sciences. For Pannenberg, theology cannot claim the universality of its truth without receiving validation by other disciplines.[1] Thus in this respect, failure to take into account the truth claims of other disciplines and other cultures would make it difficult to properly assess the validity of Pannenberg's own claims. It is precisely this type of claim to broad validity that this book aims to test, by drawing on the insights of the sciences, other disciplines, and other cultural perspectives. Can we support Pannenberg's assertion to the universal va-

1 "My truth cannot be mine alone. If I cannot in principle declare it to be truth for all—though perhaps hardly anyone else sees this—then it pitilessly ceases to be truth for me also. Theology deals with the universality of the truth of revelation"; Pannenberg, *Systematic Theology,* i. 50–1, see also 21–2. This claim to universality is a recurring one in Pannenberg's work. See e.g. Wolfhart Pannenberg, "What is a Dogmatic Statement", in id., *Basic Questions in Theology,* i (London: SCM, 1970), 182–211, esp. 201, 207; Wolfhart Pannenberg, *Jesus—God and Man,* trans. L. L. Wilkins and D. A. Priebe (2nd edn., Philadelphia: Westminster, 1968), 165; and Pannenberg, *Anthropology in Theological Perspective,* 15; among many others. For a discussion of how this approach marks Pannenberg's resistance to postmodernism see Stanley J. Grenz and John R. Franke, *Beyond Foundationalism: Shaping Theology in a Postmodern Context* (Louisville, KT: Westminster John Knox, 2001), 43–5; this leads to doubts on claims by others who see Pannenberg as a "postmodern theologian"; cf. Scott Cowdell, *Is Jesus Unique? A Study of Recent Christology* (New York: Paulist, 1996), 114–15.

lidity of his truth claims about time? Can we declare them to be "truth for all"?

Universal truth may be Pannenberg's goal, but the journey to that goal is still long and complicated, and filled with conflicting interpretations. One of the largest hazards we face is the way that the concept of time is used to shape and lend significance to that which is perceived as "historical".[2] Differing cultural and disciplinary assumptions about the nature of time lead directly to different views on what counts as "historical"—views we are likely to misunderstand if we were simply to presume that all discussions of temporal or "historical events" were based upon a universal concept of time.[3] There is an amazing breadth to the polarities which exist in various concepts of time: linear time versus circular time, objective versus subjective time, unireferential versus multireferential time. Theology adds to this another polarity, that between time and eternity (or between temporality and atemporality). Those who understand *time* as objective, linear, and excluded from transcendence will understand and, above all, seek different things from time, and evaluate historical events in very different ways from those who understand time as subjective, circular and fundamentally connected to a transcendent eternity.[4] In short, we see the immensity of the task that Pannenberg has

2 Cf. e.g. Nicolas Berdyaev, *The Meaning of History,* trans. G. Reavey (London: Bles Centenary, 1936), 63: "The nature and significance of time are the fundamental problem and premiss of every philosophy of history. ... The significance we attach to history is directly determined by that which we attribute to time."

3 For a practical example of these difficulties, we need only look at the academic and popular discussions about the historical resurrection of Jesus, which jump into debates without first considering what we, or the biblical writers, understand by "historicity". The disciples and the early church (which existed at the juncture of the Jewish and Greco-Roman worlds) operated with very different understandings of time, and thus of "the historical", from those we generally assume today.

4 This difference has been highlighted in Mircea Eliade, *The Myth of the Eternal Return* (1949; Princeton: PUP, 2005). Cf. the helpful discussion in Jürgen Moltmann, *God in Creation: An Ecological Doctrine of Creation,* trans. Margaret Kohl (London: SCM, 1997), 104–9. Furthermore, while it may be tempting to speak of different national, cultural or even religious views, we should not forget that views also develop and change *within* these social groups, not just

set for himself: to take a particular view of time and history and to have it recognized as "truth for all". Because of the breadth of this goal, and the associated difficulty of testing it, we will focus specifically on the discussion with the natural sciences.

1.2 Assessment by Coherency

To avoid introducing unnecessary methodological confusion, when assessing Pannenberg's theological work on time I will simply use the same methodological criterion for truth that Pannenberg himself uses: coherency. Pannenberg notes that

> systematic investigation and presentation itself entails also a very specific understanding of truth, namely, *truth as coherence,* as the mutual agreement of all that is true. Systematic theology ascertains the truth of Christian doctrine by investigation and presentation of its coherence as regards both the interrelation of the parts and the relation to other knowledge.[5]

There are two forms of coherency operating here. Internal coherency, regarding "the interrelation of the parts" of Pannenberg's own work, and whether they are internally consistent; and external coherency, which tests "the relation to other knowledge", or whether the claims of Pannenberg's theory correlate with what we know about the world through other disciplines. While establishing the internal coherency of Pannenberg's theories of time (that is, their internal consistency with the rest of his theology) is one important part of our task and Pannenberg's, it is not the major focus. In the fragmentation of the discussion on science and theology, theology often had little option but to content itself with narrow truth claims based on internal coherency. The so-called "two

---

across them. Niklas Luhmann rightly notes that these *inner-cultural transformations* are indeed often far more significant, especially as they lead to dramatic societal transformations and act as markers of "epochal change"; cf. Niklas Luhmann, *Die Gesellschaft der Gesellschaft,* ii (Frankfurt am Main: Suhrkamp, 1998), 997–1016.

5 Pannenberg, *Systematic Theology,* i. 21–2.

language" hypothesis—seen for example in Stephen Jay Gould's NOMA principle ("non-overlapping magisteria"), which argues that science speaks about objective facts whereas theology speaks about subjective values—largely blocked theology from testing its claims with regard to *external* coherency.[6] According to such a view, theology could speak about itself and to itself, but since its truth claims were forcefully disconnected from the natural world (the realm of the sciences), there was no way for it to test those claims against broader "reality".

Here Pannenberg's focus on external coherency in the interdisciplinary discussion between science and theology bridges this language barrier. This vital and admirable characteristic of Pannenberg's work has granted him access to the interdisciplinary discourse and a measure of respect in discussions on science and theology. His broad rejection of any kind of dualism with regard to truth, and his resoluteness to work with the "same physical reality" as the sciences, is one that places theology once more into the sphere of truth-seeking communities. Pannenberg correctly sees attempts to isolate the theological discipline from the rigorous claims of the other sciences as a reactionary move by theologians: an attempt to defend theology and the contents of faith from the increasingly powerful polemics of the natural and historical sciences by seeking refuge in an isolated or internalized "theological" sphere beyond the reach of those other disciplines.[7] In contrast, Pannenberg is eager to

6 Cf. Peters, "Science and Theology", 17–18.

7 Pannenberg claims that Bultmann, Gogarten, Kähler, Hermann, and Barth are all guilty of having taken this approach, a result of which (in Christology) is the de-historicization of the gospel and the truth of the resurrection as a historical event; cf. Wolfhart Pannenberg, "Heilsgeschehen und Geschichte I", *Kerygma und Dogma,* 5/3 (1959): 218–37 at 218; Wolfhart Pannenberg, "Kerygma und Geschichte", in R. Rendtorff and K. Koch (eds.), *Studien zur Theologie der alttestamentlichen Überlieferungen* (Neukirchen: Neukirchener, 1961), 129–40, esp. 129–34. Here we see Pannenberg's reliance on Gerhard von Rad, who stressed that such a de-historicized approach is impossible from an Old Testament perspective since "der Glaube Israels grundsätzlich geschichtstheologisch fundiert ist. Er weiß sich gegründet auf Geschichtstatsachen und weiß sich gestaltet und umgestaltet von Fakten, in denen er die Hand Jahwes wirksam sah"; Gerhard von Rad, *Theologie des Alten Testaments,* i (4th edn., Munich: Kaiser, 1965), 112, quoted by Pannenberg in "Kerygma und Geschichte", 135. "Durch

display the points of unity between theological and scientific truth and, indeed, raises the truth-claims of his theological theories in direct competition to the truth-claims of modern science.[8] This rejection of the two language hypothesis and his unification of the field of enquiry produces consequences both for theology[9] and the natural sciences.[10] Thus for Pannenberg, there can be no special "theological" time which is distinct from "scientific time", or which could not be properly universalized beyond the theological ghetto.[11] This focus on coherency pushes us to ask:

---

seine Auffassung der Geschichte Israels als Traditionsprozeß unterscheidet sich *von Rad* ... von dem üblichen Nebeneinander einer 'äußeren' und einer 'inneren' Geschichte ... Die vielleicht großartigste Leistung der traditionsgeschichtlichen Betrachtungsweise *von Rads* ist es, daß sie diese Gefahren (einer supranaturalen, unhistorischen Geschichtskonstruktion) vermeiden und trotzdem die theologische Substanz der heilsgeschichtlichen Theologie bewahren läßt"; Pannenberg, "Kerygma und Geschichte", 137.

8 Cf. Wolfhart Pannenberg, "Eternity, Time, and Space", *Zygon*, 40/1 (March 2005): 97–106, esp. 97–8.

9 "If the one God of the Christian faith is also the Lord of nature, then we will not be satisfied with this chasm"; Wolfhart Pannenberg, "Contingency and Natural Law", in id., *Toward a Theology of Nature: Essays on Science and Faith*, ed. Ted Peters (Louisville, KY: Westminster John Knox, 1993), 72–122 at 72.

10 "If the God of the Bible is the creator of the universe, then it is not possible to understand fully or even appropriately the processes of nature without any reference to that of God"; Pannenberg, "Theological Questions to Scientists", 16.

11 Pannenberg often speaks of "universality", but we need to be aware of the limits of Pannenberg's use of this term. It would be too much to expect each and every interdisciplinary discussion to involve *all* academic disciplines. Pannenberg wisely limited the number of his conversation partners to the natural sciences, philosophy and the historical sciences. Occasionally, additional insights are also drawn from the social sciences, anthropology and psychology. However, while the concept of time has also been thematized in the creative disciplines (especially in classical and modern literature as well as art), Pannenberg makes little attempt to interact with these fields. For some points of entry into this broader area of discussion in art history, see Sigfried Giedion, *Space, Time and Architecture: The Growth of a New Tradition* (Cambridge, MA: Harvard University Press, 1949), or the long-running debate in art history about the supposed link between relativity theory and the development of Parisian cubism esp. in the work of Picasso and Braque. See here Giedion (ibid. 367ff.) and

To what degree then is Pannenberg's own theory of time more satisfactory and more complete than those offered by the sciences, and can that theory cohere with our own experience of the truth of time?

1.3 One Object of Enquiry: Unified Temporality

We have seen that one of Pannenberg's advantages in his discussion with the natural sciences is his desire to do away with distinct, so-called *scientific* and *theological* times and deal instead with a single, unified reality accessible both to scientists and theologians alike. Before Pannenberg, the attempt to create a unified concept of time which did not seek to introduce a second metaphysical (or *theological*) realm, was uncommon among theologians.

One cause can be attributed to theology's continuing struggle with the long-standing (perceived) dichotomy between temporality and a-temporality, or in more theological terms between "created time" and "eternity". The theological dichotomy between eternity and temporality already introduces a hierarchical division into our thinking on time: a division between mundane, earthly time, and a higher, overarching eternity. And of course, eternity exists outside the reach of the natural sciences and is open to theology alone. Theologically, the concept of eternity as a form of divine atemporality opposed to created time has been doctrinally useful for defending against the dangers of pantheism and distinguishing the actions and nature of God from those of God's creatures. To do away with this division would be to bring these two "times" back into a single

---

Paul Laporte, who maintained that the "integration of non-Euclidean geometry with the fourth dimension is a constituent factor in contemporary physics ... [which] happened in physics at exactly the same time as the change to cubism happened in painting (Einstein, *Special Theory of Relativity*, 1905; Minkowski, 1908; Picasso's first Cubist picture, *Les Demoiselles d'Avignon*, 1906–07)". A refutation is offered by Linda Dalrymple Henderson, "Four-Dimensional Space or Space-Time? The Emergence of the Cubism-Relativity Myth in New York in the 1940s", in Michele Emmer (ed.), *The Visual Mind*, ii (Cambridge, MA: MIT, 2005), 349–98 at 350. A more detailed analysis can be found in Linda Dalrymple Henderson, *The Fourth Dimension and Non-Euclidean Geometry in Modern Art* (Princeton: Princeton University Press, 1983).

category ("omnitemporality") which would then require an awkward reworking of this relation between "time" and "eternity": either a de-temporalization and divinization of created time (which would bring it into God's eternity), or a temporalization of God's eternity which would bring it down into the profane realm of creatures.

Thomas Aquinas, for example, saw attempts to temporalize God's eternity as highly problematic, since temporality is closely associated with philosophical concepts such as potentiality, flux, and change—concepts Thomas wanted to keep separate from our understanding of God. For Thomas, since temporality necessarily involved a "before" and an "after", it implied a change of state across "times", and thus an inherent potentiality for change in objects caught up in time. Potentiality was a characteristic of *imperfect* beings, those that changed from imperfection to perfection, or devolved from perfection to imperfection—processes that Thomas could not apply to God. For Thomas, the divine and unchanging God must necessarily lack potential and thus also any temporality. If God changed, then there would be a before and an after for God, God would exist in "time", and there would be potentiality in God's nature. If this were so, then God would no longer be perfect. For Thomas "any changing thing, whatsoever the change, is somehow potential" whereas "God [is] sheerly actual and unalloyed with potentiality.... Anything in change acquires something through its change, attaining something previously not attained", whereas if God is truly perfect, then God would not only lack potentiality (the underlying precondition for change) but also any need for temporality.[12] In short, if God is eternal, uncreated, and unchanging, and if time itself is variable, created, and characterized by flux, how can both God and creation exist within the same temporal plane? In this way, eternity serves to protect the nature of God.

The early Karl Barth also critiqued theological attempts to integrate "created temporality" with "divine eternity", supporting the existence of a temporal duality by arguing that God's eternal world touches this temporal world only "as a tangent touches a circle, that is, without touching it".[13]

12 Thomas Aquinas, *Summa Theologiæ* (Blackfriars edn.; London: Eyre & Spottiswoode, 1964–81), 1a, Q9, a1.

13 For an early instance of this approach, see Karl Barth, *The Epistle to the Romans,* trans. E. C. Hoskyns (6th edn., London: OUP, 1968), 30. For a survey, see

In the forward to the second edition of his *Epistle to the Romans,* Barth stressed this division, noting that, "If I have a system, it is limited to a recognition of what Kierkegaard called 'the infinite qualitative distinction' between time and eternity".[14]

But it is becoming increasingly difficult to defend such metaphysical dualities in theology, since we have become more aware of the many difficulties they introduce.[15] For example, a division between created temporality and divine temporality would appear to push God and creation onto two distinct ontic levels of action, which would then raise the question of how God and human beings are to interact across these mutually exclusive frameworks. While there are a variety of possible solutions theologians can appeal to here, Pannenberg abandons the duality of time and eternity and instead argues that God's eternity and created temporality are coterminous, that God's eternity is simply an all-inclusive "omnitemporality", the totality of all temporal time frames. "Eternity ... does not stand in contrast to time as something that is completely different. Eternity creates no other content than time. ... Eternity is the unity of all

---

Robert W. Jenson, "Karl Barth", in David Ford (ed.), *The Modern Theologians* (2nd edn., Oxford: Blackwell, 2001), 21–36, esp. 26–8.

14 Barth, *Epistle to the Romans,* 10.

15 See e.g. Jürgen Moltmann's attempt to overcome the strict dualism between creator and creation, which also includes a reworking of the dualism between temporality and eternity, in his *God in Creation.* Or the discussions in theological anthropology aimed at overcoming the metaphysical duality in our concepts of body and soul, which seek to promote instead a unified "nonreductive physicalism"; see Warren S. Brown, "Neurobiological Embodiment of Spirituality and Soul", in Malcolm Jeeves (ed.), *From Cells to Souls—And Beyond: Changing Portraits of Human Nature* (Grand Rapids, MI: Eerdmans, 2004), 58–76, esp. 63ff.; Jaegwon Kim, *Physicalism, Or Something Near Enough* (Princeton: Princeton University Press, 2005), 33ff.; and the extended debate in W. S. Brown, N. Murphy and H. N. Malony (eds.), *Whatever Happened to the Soul? Scientific and Theological Portraits of Human Nature* (Minneapolis: Fortress, 1998); for a critical overview, see Ansgar Beckermann, "Introduction—Reductive and Nonreductive Physicalism: A Short Survey of Six Decades of Philosophical Discussion Including an Attempt to Formulate a Version of Physicalism Against the Background of C. D. Broad's Concept of Emergence", in id., Hans Flohr, Jaegwon Kim (eds.), *Emergence or Reduction? Essays on the Prospects of Nonreductive Physicalism* (Berlin: de Gruyter, 1992), 1–21.

time".[16] Thus for Pannenberg, time represents a single unified object that is open for investigation, rather than a divided system of ontic realities. Rejecting the two-language hypothesis or the ghettoization of theology and science not only means opening up science for theological access, but also opening up theology for access by scientists. Pannenberg's approach takes this bold move, refusing to hide theology and its concepts away from the investigation of the sciences. For Pannenberg, this accessible foundation in a common time is vital if we are going to have a shared, interdisciplinary discussion with the sciences at all. The question of time "is one of the most neglected but also one of the most important questions in the dialogue between theology and science. It is unavoidable if the reality of God is to be related in a positive way to the mathematical structure of the spatiotemporal world of nature".[17] And it is only by acknowledging that we have one object of enquiry open to examination by all that we can move this discussion and collaboration between theology and science forward.

1.4 Temporality as Objective; History as Subjective?

Two additional complications arise when we investigate time. First, while Pannenberg may want to overcome temporal dualisms, it has already become standard practice in much of our academic discourse to differentiate this broad field of the "temporal" into concepts of *time* and concepts of *history,* and to deal with these separately. On the one hand, this move seems fairly obvious and even necessary: there seems to be little overlap between the interests of theoretical physicists and the work of historians. The aims and intentions of physics and historiography, both in general and in their particular fields of research, appear far too divergent to justify their unification under a single academic umbrella.

But on closer examination, these disciplines differ not only in their aims and goals, but also in a way that promotes the development of a new polarization of time. What we find is that concepts of *time* (for example,

16 Wolfhart Pannenberg, *What is Man? Contemporary Anthropology in Theological Perspective,* trans. Duane A. Priebe (Philadelphia: Fortress, 1970), 74. See also Pannenberg, "Eternity, Time, and Space", 102.

17 Pannenberg, "Theological Questions to Scientists", 18.

in physics) focus on providing an overarching ontological framework, an absolute point of reference, within which the secondary course of *history* then occurs. In other words, physics and the philosophy of time seek to understand *what* time is. As to what human beings then *do with time,* this question is left to the historian and to the broader humanities. We could say that this constructed relationship between time and history reflects a Newtonian mentality, where absolute time (as a first order entity) provides a container for the performance and occurrence of physical and historical acts (which are then reduced to the level of secondary actions).[18] This Newtonian approach leads to fragmentation between *the structure of time* and its *contents,* a fragmentation that was then halted by the philosophy of relativity which recombined time and its contents into a singe four-dimensional reality (as we shall see in Chapter 3).

Second, additional complications have arisen since the end of the nineteenth century and the beginning of the twentieth century, as the focus of both concepts of time and history have slowly changed. The shift in physics to relativity theory and its views of relative frames of reference curiously paralleled a broader shift towards the appreciation in historiography of independent and relative histories.[19] Historical work,

---

18 Again, there are parallels in modern anthropology, where there is a long-standing polarity between the biological, especially genetic, sciences (which claim to represent *what* the human being really is) and the broader humanities such as theology, ethics, sociology, literature, and history, which are then left with the task of simply analysing what this "genetic conglomerate" subsequently *does.* We can see E. O. Wilson's experiments in sociobiology as one attempt to overcome this divide; cf. Edward O. Wilson, *Sociobiology: The New Synthesis* (Cambridge, MA: Belknap, 1976). However, sociobiology seeks broad unifications at the expense of a nuance and subtlety required to deal adequately with the complete data of human experience. Wilson appreciated these challenges, but hoped that advances in scientific knowledge would lead to "a genetically accurate and hence completely fair code of ethics" (ibid. 562), using the simplicity of genetics to reductively eliminate the nuance of "fairness".

19 For discussion on the possible links between the growth of relativity theory and postcolonialism, see Michael J. Watts, "Capitalisms, Crises, and Cultures 1: Notes Toward a Totality of Fragments", in Allan Pred and Michael J. Watts, *Reworking Modernity: Capitalisms and Symbolic Discontent* (New Brunswick, NJ: Rutgers University Press, 1992), 1–20, esp. 16f.

especially from new postcolonial perspectives, moved beyond the writing of a "universal history" towards an appreciation of the perspectival, personal and particularist aspects of history. This movement is now irreversible. The lessons learned in postcolonial studies have rendered any modern project to rewrite a universal history deeply questionable.[20] Yet paradoxically, the shift away from the writing of universal histories by historians has been matched by a move on the part of physical theorists back to the unification of time and its contents in a universal, relativistic (Minkowskian) notion of space-time. The rise of special relativity pushes us towards an acceptance of multiple perspectival histories, while unified space-time simultaneously renews in scientists the desire to return to universal histories with their appealing, over-arching metanarratives. John Polkinghorne expresses this shift well when he notes that the "scale of theological thinking, in both space and time, still remains domesticated and anthropocentric. When theologians speak of the 'world', they usually do not mean the universe but our local planet. When they talk of history, it is mostly the few thousand years of human cultural development that they have in mind."[21] Polkinghorne's desire that theologians redefine history in a way that is unified with the universal processes of

20 Cf. e.g. Eckhardt Fuchs, "Reshaping the World: Historiography from a Universal Perspective", in Larry E. Jones (ed.), *Crossing Boundaries: The Exclusion and Inclusion of Minorities in Germany and the United States* (New York: Berghahn, 2001), 243–63: "In Germany, it was not the status of a superpower but rather the attempt to come to terms with the nationalistic historiography after 1945 and to regain its European intellectual heritage that led to a growing interest in world history in the 1960s. Therefore, from Hans Freyer to Karl Jaspers and Alfred Heuß the Eurocentric view of world history in the vein of Ranke, Jacob Burckhardt, and Ernst Troeltsch prevailed over the German-centred view of Schäfer and Delbrück. The title of Freyer's book *Weltgeschichte Europas* shows that Troeltsch's phrase 'For us there exists only a universal history of European culture' was still the dominant credo after 1945. Without a doubt, world history in the cold war period was every bit as political and ideological as it was in nineteenth-century historiography. However, over the last two decades this has drastically changed. More recent debates on world history and postcolonialism [... have] challenged the Eurocentrism in Western historiography" (250).

21 John Polkinghorne, *Belief in God in an Age of Science* (New Haven: Yale Nota Bene, 1998), 87.

cosmic time (a direct corollary of such a unification of space-time), pulls us back towards the creation of a universal "history" at the *cosmic* level. While it has its advantages, this hope is not without its dangers. Not only does it meld time and history back into a non-nuanced, monadic entity (without an appreciation of the long-held distinction between *Historie* and *Geschichte*[22]), but it exposes an epistemological vanity. That such a demand is arising within the natural sciences highlights the degree to which the sciences still view themselves as based within an epistemological tradition of "non-perspectival objectivity", one that is both able and competent to provide a neutral and unbiased formulation of universal history based on scientific accounts of "what really happened".[23]

---

22 For the distinction in Germanic historiography between *Geschichte* and *Historie*, see e.g. the overview of Hegel's use of this distinction in Michael Inwood, *A Hegel Dictionary* (The Blackwell Philosopher Dictionaries; Oxford: Blackwell, 1992), 118–20. The distinction also quickly entered English historiography during the epistemological debates of the nineteenth century: "The word 'history' ... signifies *either* a form of literary composition *or* the appropriate subject or matter of such composition—*either* a narrative of events, *or* events which may be narrated"; R. Flint, *History of the Philosophy of History* (New York: Charles Scribner's Sons, 1894), 5, emphasis added. Not only is it impossible to free the term from this duality and ambiguity of meaning, it may also be undesirable. Under the term history we can place either those events "that really happened" (in some near-intangible "objective" or "factual" sense) or the narration of those events (in a more "subjective" sense). It is precisely this difficult and seemingly insurmountable polarity between objective reality and subjective narrative which has led a significant number of contemporary historians to view universal histories with justified suspicion, to abandon the modernist search for underlying, pseudo-scientific historical truths, and to concentrate solely on the narrative aspect of history, thus limiting "history" to the one species known "to narrate", and rejecting any sense of a "natural" or "cosmic" history. It was against this background that E. A. Freeman could raise his (in)famous claim that "history is past politics, ... politics are present history. The true subject of history, of any history that deserves the name, is man in his political capacity, man as the member of an organized society, governed according to law"; E. A. Freeman, "The Practical Bearings of European History", in id., *Lectures to American Audiences* (London: Trübner, 1882), 207.

23 This is a concern Pannenberg echoes with regard to the self-confidence of the natural sciences to determine "what really happened" in the resurrection.

1.5 Pannenberg: A Biographical Sketch

The postcolonial turn has not just made us aware of the importance of national or cultural perspectives, but also of the influential role played by individual biographies.[24] When we come to Pannenberg's work, we are lucky that attention has already been given to his particular biography and its influences on his work.[25] Rather than repeat that material here, I simply want to draw attention to a few milestones that will help create context for Pannenberg's work on time.

Born in 1928 in Stettin, Germany (now Szczecin, Poland), Wolfhart Pannenberg's family fled west during the Second World War, but then moved again in 1944 during an American bombardment on Berlin, only to be forced to move a third time by the advancing Russian offensive. On the afternoon of January 6, 1945, the 16 year-old Wolfhart Pannenberg had what he would later call his "light experience": he found himself "suddenly flooded by light" as he was walking home, "absorbed in a sea of light which, although it did not extinguish the humble awareness of my finite existence, overflowed the barriers that normally separate us from the surrounding world".[26] He would later write, "I did not know at the time that January 6 was the day of Epiphany, nor did I realize that in that

---

"The so-called methodological atheism of modern science is far from pure innocence. It is a highly ambiguous phenomenon", Pannenberg, "Theological Questions to Scientists", 16.

24 See e.g. James McClendon, Jr., *Biography as Theology: How Life Stories Can Remake Today's Theology* (Philadelphia: Trinity, 1990). See Dan R. Stiver, *The Philosophy of Religious Language: Sign Symbol and Story* (Oxford: Blackwell, 2002), 134–62 for an analysis of the importance of biography in narrative theology in the North American context. Or for culturally broader case studies, cf. Andreas Schüle (ed.), *Biography as Religious and Cultural Text* (Münster: LIT, 2002).

25 Pannenberg has also published a number of autobiographical notes, including Wolfhart Pannenberg, "An Autobiographical Sketch", in C. E. Braaten and P. Clayton (eds.), *The Theology of Wolfhart Pannenberg: Twelve American Critiques with an Autobiographical Essay and Response* (Minneapolis, MN: Augsburg, 1988), 11–18.

26 Pannenberg, "God's Presence in History", 261.

moment Jesus Christ had claimed my life as a witness to the transfiguration of this world in the illuminating power and judgement of his glory".[27]

This wartime conversion experience pushed Pannenberg towards studies in theology. He received his doctoral degree in Heidelberg in 1953 (under Edmund Schlink), and completed his *Habilitation* in 1955 with a concentration on medieval theology (specifically Duns Scotus). After accepting professorships in theology at universities in Wuppertal and Mainz, Pannenberg took up the chair for Fundamental Theology and Ecumenics, part of the Protestant Faculty of Theology at the University of Munich, where he taught until his retirement in 1994.

For Pannenberg, teaching in theology is not be confused with a focus on abstract conceptions disconnected from political life. Given the formative role of his wartime experiences, political topics played an important part in Pannenberg's theological work, and he also sought ways to apply his theology to the political world around him.[28] When asked by *The Christian Century* in 1981 to describe how his theological thinking had changed over the course of his career, Pannenberg could only offer one concrete example (apart from general statements regarding methodological developments, and his teenage conversion experience): his self-described suspicion of the "liberal democratic tradition". He wrote that:

27 Ibid.

28 Cf. e.g. the collection of presentations in Wolfhart Pannenberg, *Human Nature, Election and History* (Philadelphia: Westminster, 1977) later published in German as *Die Bestimmung des Menschen: Menschsein, Erwählung und Geschichte* (Göttingen: Vandenhoeck & Ruprecht, 1978); or Wolfhart Pannenberg, *Theology and the Kingdom of God* (Philadelphia: Westminster, 1969), and further political writings such as Wolfhart Pannenberg, "Reich Gottes und Nationalismus. Vom politishen Sinn der christlichen Hoffnung", in H. J. Schultz (ed.), *Kontexte*, i (Stuttgart: Kreuz, 1965), 41–8; Wolfhart Pannenberg, "Das auserwählte Volk oder: Wird die Nation zum Götzen?", *Deutsches Allgemeines Sonntagsblatt*, 27 (4 July 1976), 26; Wolfhart Pannenberg, "Reich Gottes in Amerika. Religiöse Selbstdeutung als Thema politischer Theologie", *Evangelische Kommentare*, 10 (1977): 333–6; and Wolfhart Pannenberg, "Die Aufgabe einer politischen Theologie des Christentums", in H.-W. Bartsch (ed.), *Kerygma und Mythos VII–1. Glaube und Politik—Religion und Staat. Zur Entmythologisierung und Neubestimmung ihres Verhältnisses* (Hamburg: Reich & Heidrich, 1979), 19–25.

> The course of the students' revolution in Europe, especially the unexpected susceptibility to Marxism on the part of many educated youth, made me more keenly aware of the unpredictability of irrational factors still shaping the course of history. ... However, the more insecure the future of a liberal, secular society appears to be, the more confident I feel about the future of religion. ... It will outlive every ideological regime.[29]

In his introduction to an early colloquium on Pannenberg's work on history, James M. Robinson already identified the impact of the war on Pannenberg's thought.[30] Just as *De civitate dei* reflected Augustine's experience of the fall of the Roman Empire and his search for meaning in the aftermath[31] so too, Robinson suggests, is the impact of war and the questions of divine providence and control evident on the style of theology and the interpretation of time and history that Pannenberg offers. In a world that is being torn apart by hot and cold wars, how do we understand the guiding hand of God in history, and God's providential care for those who exist—and suffer—in time?

---

29 Wolfhart Pannenberg, "God's Presence in History", 263.

30 Robinson, "Revelation as Word and as History", 3: "if it is true that reflection upon history emerges in the wake of great historic events, then, conversely, the ways in which such events are themselves experienced should foreshadow the understanding of history operative in the historical study that follows."

31 Cf. Hans von Campenhausen, "Augustin und der Fall von Rom", in id., *Weltgeschichte und Gottesgericht: zwei akademische Vorträge über Augustine und Luther* (Stuttgart: Kreuz, 1947), 2–18.

CHAPTER TWO

CATEGORIZING TEMPORAL SYSTEMS

One of the challenges we face when we talk about the nature of time is the difficulty of navigating the rich variety of concepts, each of which reflects a particular cultural context and social background, each with its own assumptions and conclusions about the way time functions, and the way we exist in time. Without being able to enter into a detailed examination of each such temporal system, it will still be helpful to develop a simple method of classification for these different models. Creating some simple signposts for the maze of temporal systems—highlighting overarching characteristics, areas of affinity, as well as points of difference—will help us better grasp the concepts we find in Pannenberg's work and in the natural sciences.

2.1 Linearity, Objectivity, Reference

At a very simple level, we can divide theories of time roughly into cyclical and linear conceptions. As the names suggest, cyclical concepts of time focus on the *circularity* of time, while linear concepts stress the *vectorial* aspect of time, that is, the "arrow of time" which plots out contingent events from past to future. Mircea Eliade has shown the extent to which the cultures in the early history of humanity were marked by a focus on the eternal, circular return of primordial time into present life. The primordial events of mythic life are held up as the truly real, and are sought out in the continuing cycle of time. The contingent and novel aspects of historical experience are meaningless and to be threshed away in order to find and then preserve the repeating kernels of eternity and mythic significance. It is only the repetition of mythic events, the eternal return of primordial time in the present that gives times its meaning.[1]

1 "With each periodical festival, the participants find the same sacred

While there are hints of such circular time in the history of European thought (particularly in the recurring festivals of Jewish and Christian religious life), linear views of time, and the concept of history to which it leads, have come to dominate. In the modern period, there have been some attempts to recover aspects of a cyclical understanding of time, with Oswald Spengler and Giambattista Vico each offering their own examples.[2] Yet to take Spengler for example, we see that his system is not cyclical in the same sense as Eliade's eternal return. Rather than focusing on the *reactualization* of primordial, mythic time through the cult, Spengler is more interested in the narrower view of individual cultural morphology. What is important for Spengler is simply the "biological" development of a culture in analogy with other living organisms: the movement from birth to adolescence and maturity and then, via decadence, into decay and death.[3] For Spengler, time is cyclical only in the same way that a progressing collection of biographies can be said to be cyclical, in that they all display the same repetitive (yet still discrete and independent) biological development from birth to maturity and then death. The difficulty of finding truly circular conceptions of time in the modern West highlights the degree to which Western thought on time has come to be characterized by overwhelmingly linear theories of time, and dominated by the linear theorists and champions of progress,

---

time—the same that had been manifested in the festival of the previous year or in the festival of a century earlier; it is the time that was created and sanctified by the gods at the period of their *gesta,* of which the festival is precisely a reactualization. ... Hence religious man lives in two kinds of time, of which the more important, sacred time, appears under the paradoxical aspect of a circular time, reversible and recoverable, a sort of eternal mythical present that is periodically reintegrated by means of rites. This attitude in regard to time suffices to distinguish religious from nonreligious man; the former refuses to live solely in what, in modern terms, is called the historical present"; Mircea Eliade, *The Sacred and the Profane: The Nature of Religion* (Orlando, FL: Harcourt, 1987), 69–70.

2 For a broader discussion of Spengler, Vico and cyclical time, including reference to the affinity between cyclical theories of time and Buddhism, see David Bebbington, *Patterns in History: A Christian Perspective on Historical Thought* (Vancouver: Regent College Publishing, 1990), 21–42.

3 Cf. Oswald Spengler, *Decline of the West,* trans. C. F. Atkinson (Oxford: Oxford University Press, 1991), 17–18.

such as Hegel, Marx and (importantly for the natural sciences) Darwin.[4] These beliefs in cultural and biological progress find a deep root in the underlying conviction that time has a real flow forward and (despite its detours and diversions) is linear in character.

Of course, such a classification involves a degree of simplification; few systems can be cleanly categorized as purely linear or purely cyclical. On the one hand, Buddhist models of time, despite their general circularity, can still display a linear, eschatological and apocalyptic character.[5] On the other hand, the discussion of time in modern physics also exhibits aspects of circularity. Physics may swap out simple linearity for the vector of "entropy" (which seeks to provide time with an arrow, and with basic directionality), and yet this vector is not always clearly linear, directional, or measurable in all cases. This leads Stephen Hawking to speak of "three different arrows of time": a thermodynamic arrow (based on entropy), a psychological arrow (based on our subjective perceptions), and a cosmological arrow ("the direction of time in which the universe is expanding rather than contracting").[6] The subsequent directionality of time is far from assured unless one is able to use more than one measure simultaneously. For example, the second law of thermodynamics may provide a neat division between before and after in simple systems which move from organization to randomness, but once an isolated system reaches thermodynamic equilibrium (or a state not easily distinguishable from it), this arrow of time is lost. After all, it is easy to distinguish the *before* and *after* states if we see a picture of vase and a second picture of it shattered. However, if the shards are then re-broken or randomly rearranged and then re-photographed on successive days, it would be far more dif-

4 Given the strong dialectical characteristics of their models, it may seem odd to think of Hegel and Marx as "linear theorists", yet despite their dialectical systems they still display the general conviction that time, despite its tensions, "progresses".

5 Cf. H. Seiwert, "End of Time and New Time in Medieval Chinese Buddhism", in A. I. Baumgarten (ed.), *Apocalyptic Time* (Leiden: Brill, 2000), 1–14. For a practical discussion, cf. John S. Strong, *Relics of the Buddha* (Princeton: PUP, 2004), 211–28.

6 Stephen Hawking, *A Brief History of Time: From the Big Bang to Black Holes* (London: Bantam, 1988), 145.

ficult to place these new photos into chronological order.[7] Without recourse to an external system (either with its own measurable entropy, or with an appeal to the "psychological" or "cosmological" arrows of time), this one compass arrow of time begins to spin. The linear arrow of time in physics is then complicated further if we postulate the cyclical contraction of the universe into a "big crunch" with a subsequent new "big bang". Here we find physics beginning to develop particular affinities to circular concepts of time.[8]

Having distinguished between linear and cyclical times, we can then move to a second level, dividing between objective and subjective subcategories.[9] An example of objective time can be seen in the work of Isaac Newton. For Newton, time was the absolute container in which the universe existed and moved, and thus it was independent of the physical universe. But most importantly, this form of time could be called objective because it existed external to and independent of human consciousness. Today, the physical sciences remain the major proponent of objective models time, which find their culmination in Minkowski's interpretation of relativistic space-time.[10] In contrast, whereas objective time has little

7 Using pictures of the event removes the possibility of appealing to Hawking's "psychological" arrow of time, since the subject is removed from the event itself. In fact, Hawking later unifies the thermodynamic and psychological arrows arguing for their agreement with the cosmological arrow, though he seems to imply the final primacy and importance of the thermodynamic principle over the other two (Hawking, *A Brief History of Time*, 152).

8 This is already occurring (though at a very elementary level) in the science and Buddhism discourse. See e.g. the discussion between experimental physicist Anton Zeilinger and the 14th Dalai Lama, in Anton Zeilinger, "Encounters Between Buddhist and Quantum Epistemologies", in B. Alan Wallace (ed.), *Buddhism and Science: Breaking New Ground* (New York: Columbia University Press, 2003), 387–97, esp. 396–7, though Zellinger does not attempt to systematize the results of the discussion.

9 Indeed the division of time into objective and subjective categories has become so commonplace that Reinhart Koselleck can simply refer to it offhandedly as a given in the blurb to his book *Vergangene Zukunft*: "In general terms, we often like to distinguish between a subjective and objective time"; Reinhart Koselleck, *Vergangene Zukunft: Zur Semantik geschichtlicher Zeiten* (Frankfurt am Main: Suhrkamp, 1989), 2.

10 See the extended discussion in the following chapter. Again, note that there

interest in the workings of human consciousness, subjective views of time are much more interested in the influence of human experience and consciousness on time, and often take as their starting point Augustine's classical, ground-breaking work on the topic.[11] Rather than viewing time as existing properly outside the scope of human consciousness, subjective views of time see the experience of temporality inherently linked to our present experiences and past remembrances, and our future hopes and expectations. Time has meaning only as a signifier for the subjective mind that perceives it.[12]

Finally, at a tertiary level, we can then further subdivide these objective and subjective theories into multireferential and unireferential groups. While multireferential systems are characterized by a pluralistic matrix of multiple, equally-valid frames of reference, unireferential systems argue for the existence of a *single* universally valid frame of reference.

Describing temporal theories in this way offers us three simple spectra upon which we can place concepts of time and history: linearity–circularity, objectivity–subjectivity, and multireferentiality–unireferentiality. In the following two chapters, we will see that the concepts of time put forward by relativity physics and by Wolfhart Pannenberg actually share a great deal in common, since both can be classified as linear and objective. Where they differ is with regard to referentiality, and the power that stands behind and drives that system of referentiality.

---

is a particular unavoidable simplification at work here: not all systems can be easily divided into subjective or objective camps. With its fragmentation across multiple, individual frames of reference, we could argue that special relativity and Minkowski space-time also present aspects of a subjective system. However, the presentation of Minkowski space-time in the next chapter will argue that this model properly belongs within objective systems of time since space-time manifold (as a four-dimensional, invariant object) incorporates, transcends, and is external to human consciousness of it.

11 Cf. esp. Augustine, *Confessiones,* XI, esp. 20. For a treatment, see Moltmann, *God in Creation,* 112–18.

12 In the modern period, the models developed by Reinhart Koselleck, A. M. K. Müller, and Jürgen Moltmann are extensions of this subjective model. We will discuss this in greater detail below.

2.2 Can Theology Still Speak About Time?

At least since the 1920s, Einstein's theory of relativity has come to represent the dominant theory of time within schools of physics in the West. Yet we can see the immense power of Einstein's theory in the way that it has been absorbed into a range of other diverse academic discourses outside the natural sciences, particularly into philosophy and the social sciences, though also into theology. This movement is surprising since the theoretical language of relativity is abstract mathematics, and mathematics is not always the most appropriate language for interdisciplinary discourse (especially with the human and theological sciences). Yet the power of Einstein's work has pushed the theory of relativity beyond the walls of the university and is now firmly planted in society's imagination. The influence is so great that a mathematically-uneducated lay person may not understand anything of Einstein's theories or the scientific resources that underly them, but nevertheless live quite comfortably with the belief that the theory of relativity represents the *de facto* "academic" interpretation of the concept of time.[13] What is exceptionally impressive is that relativity theory has become this standard despite its heavy basis in abstract mathematics and its theoretical inaccessibility, despite what we might call its transcendent mysteriousness.

It was this transcendent mysteriousness, and its power, that irritated Alfred North Whitehead, one of relativity's earliest critics. Whitehead was particularly concerned that relativity theory moved physics away from its powerful grounding in common sense. For Whitehead, the great stability of classical physics came from its support of common sense, its support of a science grounded "upon what every plain man could see with his own eyes".[14] In the classical and Renaissance periods, science was limited to the observation and systematization of ordinary experi-

13 "In Western cultures, public expectation relies predominantly on scientific procedures and not much, if indeed at all, on theology or the human sciences in the endeavour to come close to the truth and evaluate evidence"; John Polkinghorne and Michael Welker, "Introduction: Science and Theology on the End of the World and the Ends of God", in eid. (eds.), *The End of the World and the Ends of God: Science and Theology on Eschatology* (Harrisburg, PA: Trinity, 2000), 1–16 at 6.

14 A. N. Whitehead, *Science and the Modern World* (New York: Free Press, 1967), 114.

ences and ideas familiar in ordinary life.[15] Suddenly, at the turn of the twentieth century—and in the 1904 to 1925 watershed period in physics[16]—Whitehead witnessed a shift away from common sense onto "a higher imaginative level"[17] where the component concepts of physics, such as ether, waves, interference, and the motion of the earth were remote from ordinary experience.[18] Whitehead had already noted that in the past "the mentality of an epoch springs from the view of the world which is, in fact, dominant in the educated sections of the communities in question"[19], but there was now never such a vast difference between the conceptions held by "the educated" and "the masses" as there was at the beginning of the twentieth century. And so with the introduction of the physics of relativity, Whitehead sees a contemporary divergence in this mentality: in the period of classical physics, the "common man" and the physicist shared the same presumptions regarding the physical world, since they were both operating with common sense. Yet today, while the "common man" is still reliant on common sense, physicists recognize the insufficiency of "ordinary observation". Common sense, by definition, is common and general, whereas the new physics of relativity is, in contrast, limited and highly specialized and has very few points of contact with everyday experience. Whitehead saw the move into relativity as a direct challenge to common sense, since pre-relativity physics "had only refined upon the ordinary notions of ordinary people".[20] And thus suddenly the dominant ascendency of relativity theory was complete: it removed knowledge of the structures of reality from the realm of common

15 Ibid. 115.

16 Opening with Poincaré's first insights into what he termed the "principle of relativity", through Einstein's development of the special theory of relativity, the general theory of relativity, Minkowski's interpretation and unification of Einstein's work, the development of quantum theory and the so-called "New Quantum Theory" introduced by Heisenberg in Autumn 1925.

17 Whitehead, *Science and the Modern World,* 114.

18 Ibid. 116. Interestingly, rather than seeing this shift as grounded upon the "finer imagination" of physicists, Whitehead attributes this development primarily to improvements in instrumentation and the increasing impetus of competitive nationalism (ibid. 114).

19 Ibid., preface.

20 Ibid. 116.

sense, and by doing so it removed its theories and its findings from the realm of public discussion and critique. Here now was a new religious orthodoxy, a transcendent mysteriousness that one could only accept and believe on the basis of the teachings of an elite priesthood. Suddenly, physics had moved from offering detailed analyses of commonly accepted descriptions of the universe to creating a transcendent reality disconnected from the common perceptions of the "common man", solidifying the bifurcation in the discourse on reality maintained still today in the "two-language hypothesis" of the science and theology debate.

Given this *de facto* supremacy of the natural sciences in regard to theories of time and the deep foundations of the discourse in cultic language of abstract mathematics, we must ask quite soberly if theology is able to enter into this discussion, or if there really is any room in the public sphere to develop or even suggest a competing theory? If the common assumption of Western societies is that "the measure and model for truth claims is found in the sciences"[21], is it at all possible to make space in the discourse for a dissenting, non-scientific voice on time? This is not to suggest that there are no dissenting voices[22], but that in the West those voices shelter in the safe harbour of the "two-language hypothesis", seeking mutual coexistence rather than directly confronting the dominant position of scientific interpretation. Direct confrontation risks being immediately (and often justifiably) dismissed out of hand.[23] Wolfhart

21 Polkinghorne and Welker, "Introduction", 6.

22 Whitehead's own process philosophy has grown to become one of the most significant alternatives, though its level of traction outside its own philosophical schools remains limited.

23 What I am not arguing for here is a type of social constructivism that subsumes the natural sciences under a religious, political or cultural ideology. Rather I am simply pointing to the openness of the discourse in many non-Western contexts. Many Chinese as well as Islamic cultural contexts (e.g. Pakistan, Egypt, Saudi Arabia) do not display the same social dominance of the sciences. This is often driven by a suspicion of the dimensions of cultural dominance represented by scientific (and often specifically relativity) theory and its close associations with Christian Western culture and Christian Western academic traditions. In Pakistan, relativity theory has been seen as an affront to the absolute knowledge provided by the Koran, and in China the Chinese Communist Party has condemned Einstein's theory of relativity as "Western bourgeoisie physics" that leads

Pannenberg's work on time represents just such an approach of direct confrontation—yet not one that calls into question the entire practice of scientific enquiry.

Throughout Pannenberg's extensive work within the science and theology discourse, what has marked his approach as distinctive from others has been his refusal to allow the physical sciences in particular—and the broader natural sciences in general—to dominate exclusively all conceptual understandings of the nature of time. Rather than submit to the dominance of certain schools of physics which claim relativistic space-time (with its mathematical and geometric expression in the Minkowski universe) as the only competent and defensible understanding of time for a modern academic discourse, Pannenberg has been rigorous in his attempts to keep the discussion of time open to analysis by the broader academy, and to defend the right of those within the humanities (including the schools of philosophy, history, and in particular theology) to present a detailed critique of the scientific conceptions of time.

> The concepts of space and time are important not only in physics and in geometry but in all human experience. It is not self-evident that the definition of these concepts is an exclusive prerogative of geometry and physics. Certainly, the measurement of spatial and temporal relations is a matter of special competence of geometers and physicists, but it is by no means certain that the measurement of spatial and temporal relations exhausts the concepts of space and time.[24]

---

to the promotion of liberal and pluralistic social values; cf. the detailed critique by Meera Nanda "The Epistemic Charity of the Social Constructivist Critics of Science and Why the Third World Should Refuse the Offer", in Noretta Koertge (ed.), *A House Built on Sand: Exposing Postmodernist Myths About Science* (Oxford: Oxford University Press, 1998), 286–308 at 294–5. This option is closed to us in the West. The vigorous critique directed at the German philosopher Karen Gloy for her statements that "Nature as such does not exist. There are only interpretations", and that nature is "a culturally specific product", show the dangers and limitations of attempting interdisciplinary discussion with the sciences in such a way. Karen Gloy, "Die Geschichte des Naturverständnisses", presentation at the 7th Ökumenischen Sommerakadamie 2005, Linz, 14 July 2005, or her longer treatment: Karen Gloy, *Das Verständnis der Natur* (München: C. H. Beck, 1995).

24 Pannenberg, "Eternity, Time, and Space", 97–8.

The dominance of space-time relativity in modern, Western discussions of time has led to an unwillingness among representatives of the humanities to formulate varying, competing approaches to time. Pannenberg rightly understands this situation as particularly problematic for theology, since it leaves us with a poor choice. On the one hand, theology can choose to establish a metaphysical duality, developing its own theory of time in separation from the natural sciences, but thus be relegated to a theological ghetto of the "supra-historical" and "supra-temporal", disconnected from the scientific discussion of the reality of this world.[25] Or on the other hand, theology might accept that the sciences offer a totally sufficient account of the nature of the world and time, and agree to be bound by the limits of that theory—but then be forced to alter or abandon what it feels are the truth claims of the Christian faith (such as the historical, bodily resurrection of Jesus[26]). Pannenberg sees both options as insufficient.

If theology testifies to a God who is known in history and revealed within time[27], then theology cannot hide away in a supra-historical "theological reality". Yet if theology wants to maintain its own truth claims in contradistinction to the truth claims of the natural sciences, then it cannot agree to be bound by arbitrary limits of scientific theory or to have the validity of its truth claims dictated to it by such theories, especially if "the one God of the Christian faith is also the Lord of nature".[28] If we succumb to the rules of the two language hypothesis—that science speaks the lan-

---

25 An approach Pannenberg sees taken by Kähler and Barth; cf. Pannenberg, "Heilsgeschehen und Geschichte I", 218.

26 Defending the reality of the resurrection is a particular concern for Pannenberg in the science and theology debate. Cf. Wolfhart Pannenberg, "Heilsgeschehen und Geschichte (Fortsetzung und Schluß)", *Kerygma und Dogma*, 5 (1959): 259–88, esp. 263, 266; Wolfhart Pannenberg, "Did Jesus Really Rise From the Dead?", *Dialog*, 4 (1965): 128–35; Wolfhart Pannenberg, "History and the Reality of the Resurrection", in Gavin D'Costa (ed.), *Resurrection Reconsidered* (Oxford: Oneworld, 1996), 62–72; and Pannenberg, *Systematic Theology*, ii. 358–9.

27 "The fundamental proof for the divinity of YHWH exists in his acts in history"; Wolfhart Pannenberg, "Introduction", in id. (ed.), *Revelation as History*, trans. D. Granskou (Toronto: Macmillan, 1969), 3–21 at 13.

28 Pannenberg, "Contingency and Natural Law", 72.

guage of fact while theology speaks the language of value—then theology can maintain its truth claims in the presence of the natural sciences only if their respective realities remain strictly separated.[29] Yet Pannenberg rejects this proposed relationship between theology and the sciences[30], rejects these two options, and instead follows a third: to demand rigorous theological truths within a shared reality, to which both science and theology can testify. Such an approach can function due to the tight overlap in Pannenberg's work between ontology and epistemology, between a conception of God as creator and our ability to understand the natural world as that creation. Pannenberg argues that "If the God of the Bible is the creator of the universe, then it is not possible to understand fully or even appropriately the processes of nature without any reference to that God".[31] Of course, in the end this remains a theological claim. The challenge that faces Pannenberg is to display that the natural world indeed *cannot* be understood fully without God, or in other words that secular theoretical physics alone is not enough to understand the nature of reality. As a statement of faith, based on faith in the Creator God, the onus is on Pannenberg (and in broader terms, on theology) to support this position, and to reintroduce space for a theological perspective into the discussion of time by displaying this claimed general insufficiency of scientific theory in explaining the totality of the world about us. While Pannenberg can certainly claim this insufficiency, the claim must be supported.

But can theology achieve this? Can we create space in the physical theory of time for a more nuanced understanding of temporality? The conception of time put forward by relativity physics has in many circles become so dominant, and so unquestioned in its dominance, that it has

---

29 According to the two language hypothesis, both fields deal with reality but are nevertheless separated in their respective discussions and, if they hold to discussing their proper subject matter, should avoid not only confrontation but also interaction. See e.g. Ted Peters, "Naturwissenschaft und Religion", 57–8.

30 The natural sciences and Christian faith cannot ignore each other or operate in mutually exclusive worlds, "because they are both concerned with the same world"; Wolfhart Pannenberg, "Theologie der Schöpfung und Naturwissenschaft", presentation at the 7th Ökumenischen Sommerakadamie 2005, Linz, 13 July, 2005.

31 Pannenberg, "Theological Questions to Scientists", 16. Cf. also Pannenberg, "Eternity, Time, and Space", 97.

detrimentally affected not only the opportunity for postulating serious alternatives, but even the simple plausibility of critique. Pannenberg, in developing his own independent theory, has recognized the preliminary need to present an adequate critique of scientific views of time, simply in order to open up a theoretical space where other conceptions can then be developed and considered. With this aim in mind, the following chapter will provide a brief sketch of the development of Minkowski space-time, highlight the aspects of the theory that raise difficulties from a theological perspective, and present Pannenberg's points of critique regarding the (in)sufficiency of objective, relativistic space-time.

2.3 Subjective Theories of Time

Before we launch into a critique of relativity theory, and attempt to make space in the discussion for a theological concept of time, it will be helpful to examine closer certain aspects of subjective theories of time. Although the proponents of subjective models of time appear less vocal, and their concepts have yet to gain widespread acceptance or traction in the Western discussion, it would be unfair to continue on without saying something about the opposition and critique they also offer to the objective models put forward by the sciences. The most important subjective models can be found in Augustine's classical work, and in the modern scholars (such as Reinhart Koselleck) who have developed his heritage. The Russian philosopher of religion Nicolas Berdyaev also offers us a useful "mixed form" that bridges the subjective and objective camps. These quick portraits will provide better contrast for our coming discussion of the objective systems developed by Pannenberg and the natural sciences.

2.3.1 Augustine and his Heritage

Augustine's famous and often quoted discourse on time in Book XI of his *Confessions* approaches the topic with a focus upon our subjective perceptions of the phenomenal world.[32] The initial starting point for Augus-

32 See esp. XI, 14, 18 and 20.

tine's reflection is the ontological basis of time. Whereas Aristotle also began his discussion of time at the ontological level[33], Augustine specifically addresses the apparent ontic *absence* of that which is past and future:

> I say with confidence that I know that if nothing passed away, there would be no past time; and if nothing were still coming, there would be no future time; and if there were nothing at all, there would be no present time. But, then, how is it that there are the two times, past and future, when even the past is now no longer and the future is now not yet? But if the present were always present, and did not pass into past time, it obviously would not be time but eternity. If, then, time present—if it be time—comes into existence only because it passes into time past, how can we say that even this is, since the cause of its being is that it will cease to be?[34]

In tension here for Augustine is the apparent shift of objects from non-existence in the future, into existence in the present and then out of existence again in the past. When perceived in this way, the phenomena of time suggest that "time is only as it tends toward nonbeing"[35], with the present forming an infinitely small sliver of being between the nonbeing of the past and the nonbeing of the future.

Therefore, if "time is only as it tends toward nonbeing", how are we to understand the nature of those objects and events which now "no longer" are (past) or "not yet" are (future)? It is at this point that Augustine appeals to the subjective human mind in the present as that which gives meaning to the temporal concepts of past and future via subjective human memory and expectation.

> If there are times past and future, I wish to know where they are. But if I have not yet succeeded in this, I still know that wherever they are, they are not there as future or past, but as present. For if they are there as future, they are there as "not yet"; if they are there as past, they are there as "no longer". Wherever they are and whatever they are *they exist therefore*

33 Cf. Aristotle, *Physics*, IV, 219a.

34 Augustine, *Confessions*, XI, 14. English translations taken from Augustine, *Confessions and Enchiridion*, trans./ed. A. C. Outler. (Philadelphia: Westminster, 1955).

35 Ibid.

> *only as present.* Although we tell of past things as true, they are drawn out of the memory—*not the things themselves, which have already passed,* but words constructed from the images of the *perceptions which were formed in the mind,* like footprints in their passage through the senses. My childhood, for instance, which is no longer, still exists in time past, which does not now exist. But when I call to mind its image and speak of it, I see it in the present because it is still in my memory. Whether there is a similar explanation for the foretelling of future events—that is, of the images of things which are not yet seen as if they were already existing—I confess, O my God, I do not know. But this I certainly do know: that we generally think ahead about our future actions, and this premeditation is in time *present;* but that the action which we premeditate is not yet, because it is still future. When we shall have started the action and have begun to do what we were premeditating, then that action will be in time present, because then it is no longer in time future.[36]

The key here is Augustine's rejection of the ontic stability of these events: the past and future "exist ... only as present", and in the present exist only as "perceptions ... formed in the mind". They have no independent ontological basis outside of the mind, outside of the individual's own capacities for remembering those events which have now ceased to be. Once at this point, Augustine is free to apply this approach in a comprehensive way in developing his subjective concept of time.

> It is manifest and clear that there are neither times future nor times past. Thus it is not properly said that there are three times, past, present, and future. Perhaps it might be said rightly that there are three times: a time present of things past; a time present of things present; and a time present of things future. For these three do coexist somehow in the soul, for otherwise I could not see them. The time present of things past is memory [*memoria*]; the time present of things present is direct experience [*contuitus*]; the time present of things future is expectation [*expectatio*].[37]

Thus rather than speaking simply of past, present, and future as objective realities, Augustine argues that one can rightly only speak of a "time present" (as the only state with any ontic basis), and divide this "time

36 Augustine, *Confessions,* XI, 18, emphasis added.
37 Ibid. XI, 20.

present" into three directional perspectives: one directed at past things, one at present (existent) things, and one at future things. As such, the divided nature of time finds its basis within the capability of the human mind to direct its attention in these three ways: through memory, direct experience, and expectation.

The impetus behind Augustine's search for a new understanding of time was the problem of the lasting reality of objects. Can we attribute ontic reality to the unseen objects and events of the past and future, or are we left with an ontology which sees objects suddenly popping into present reality from the future, only to pop out again into the past? As we will see in the next chapter, one of the great achievements of Minkowski space-time was in addressing this very question, providing both the past and future with an objective, thoroughgoing ontological stability within the static and unchanging four-dimensional universe. While seeking the same ends, Augustine's approach laid the basis for this stability instead in the subjective nature of human (and finally divine) consciousness, an approach which would have a long-lasting effect within the discourse on time.

In the modern discussion, Reinhart Koselleck picked up Augustine's approach and expanded on it in his own work on experience (*Erfahrung*) and the future (*Zukunft*). As with Augustine, Koselleck avoids discussing the concepts of past, present, or future without also considering the influence of both the individual and communal subjects, who evaluate and constantly *re*-evaluate the objects of experience. He offers the following definitions:

> Experience [*Erfahrung*] is *present past,* whose events have been embodied and can thus be remembered. Experience includes a rational reworking, together with unconscious modes of behaviour which are not (or no longer need to be) present in awareness. Furthermore, one's own experience always contains and preserves foreign experiences that have been adopted through generations or institutions. ... Similarly with expectation [*Erwartung*]: it too is *bound to a person, yet is also interpersonal;* expectation also takes place today. It is the *future made present.*[38]

38 Koselleck, *Vergangene Zukunft*, 354–5.

For Koselleck, as with Augustine, the past and future are both actualized in the present, "made present" (*vergegenwärtigt*) through the individual or collective experience of the person.[39] Yet Koselleck's subjective concept of time extends beyond Augustine's in two ways. First, Koselleck's conception is more communal (or multireferential) in nature compared to Augustine's more unireferential approach. For Koselleck, temporal experiences are still made present and bound to an individual, and yet they are also formed and transmitted in an inter-generational, institutional, *interpersonal* way. Second, since experiences are inherently connected with subjective perceptions, experiences are by no means static entities. While the underlying data of the experienced phenomena may in some sense be stable, that data is subject to constant reappraisal and reassessment by the individual and communal subject, being reworked in light of future experiences, hopes, expectations, and disappointments.[40]

It is precisely this interconnectedness of experience with subjective perception and its re-evaluation dependent upon one's perspective which A. M. K. Müller picks up in his own matrix of modalized time. Whereas Augustine operated with a three phase unidimensional concept of subjective time, centred in the present and looking back (upon what was past) and forward (to what is expected in the future), Müller expands this basic position into nine phases across three dimensions: the past of the past, the past of the present, the past of the future; the present of the past, the present of the present, the present of the future; the future of the past, the future of the present, and finally the future of the future, where

39 Yet the future, though made present like the past, is not linked with experience in the same way. The future can be expected (*erwartet*) in the present, but if it were already experienced, it would rob the future of its surprising nature. The "future made present ... aims at the not-yet, at that which has not been experienced, at that which can only be inferred"; Koselleck, *Vergangene Zukunft*, 355. "If they are grounded upon experience, expectations, when they are fulfilled, will no longer surprise us. Only that which is not expected can surprise us, for then there is a new experience. Thus the penetration of the horizon of experience promotes new experiences. This newly-gained experience then exceeds that possible future which was limited by our previous experiences. ... It is the tension between experience and expectation which, in various ways, pushes us toward new solutions and in this way drives forward historical time"; ibid. 358.
40 Ibid.

each phase reflects a differing subjective disposition towards the content of our temporal experiences and expectations. The incredible complexity here is thoroughly intentional. With his "matrix of modalized time", Müller purposefully sets out to destroy the "aura of unidimensionality" of objective interpretations of the present.[41] Yet while Müller certainly succeeds in adding complexity to the discourse, it comes at a high price. The matrix purposefully aims to fragment the interdisciplinary discourse on time, relegating differing disciplines to different cells of the matrix. For Müller, the sciences and fine arts are consigned to the past present, present present, and future present (since they deal with "phenomenal time"), whereas history and religion (which deal with "non-phenomenal time") are relegated to the past past, the past future, the future past and future future.[42] Müller's matrix pushes the subjective theory of time to its most extreme, and yet it leaves us with a highly fragmented and segregated understanding of time. In the decades since its publication, Müller's subjective concept of time has failed to gain significant traction within the broader discussion on time, or to make much of an impact on the dominance of objective models.

2.3.2 A Mixed Model: Nicolas Berdyaev

Unlike Koselleck and Müller, the Russian philosopher of religion Nicolas Berdyaev offers a concept of time that impacts more directly on theological understandings of temporality, and he does this by mixing aspects of both objective and subjective models of time. On the one hand, Berdyaev seeks to defend a radical subjectivity (within a philosophy of personalism), while also wanting to accept objectivized forms of time. Thus Berdyaev's system[43] displays a unique blend of circularity and lin-

41 A. M. K. Müller, *Das unbekannte Land: Konflikt-Fall Natur; Erfahrungen und Visionen im Horizont der offenen Zeit* (Stuttgart: Radius, 1987), 218.

42 Ibid.

43 Across the many years of his work, Berdyaev offered several differing studies of time, beginning in his volume *The Meaning of History,* trans. G. Reavey (London: Geoffrey Bles, 1936), and developed further in his later writings in *Solitude and Society,* trans. G. Reavey (London: Geoffrey Bles, 1947) and *Slavery and Freedom,* trans. R. M. French (New York: Charles Scribner's Sons, 1944). In

earity, objectivity and subjectivity, as well as individualism (unireferentiality) and collectivism (multireferentiality). In his later period, Berdyaev outlines what he calls "the three kinds of time"[44]: cosmic time, historical time, and existential time.[45]

Cosmic time "is nature's time" and is symbolized by the circle and by "circular movement": "the motion of the earth around the sun, ... the reckoning of days, months, and years, ... the calendar and the clock ... [where] a return is constantly taking place, morning comes and evening, spring and autumn".[46] As a "rhythmic", circular form of time, cosmic time allows for changes not only *in* time, but the change *of* time itself, for "a turning back of time". Berdyaev identifies cosmic, natural time as the most elementary form of time, seeing it connected with the existence and movement of objects. Cosmic time "exists because movement and change take place", therefore it is *objectivized* time, "subject to mathematical calculation; ... subject to number, to division into parts and to aggregation".[47]

Yet for Berdyaev, this circular, objectivized time is inherently fallen and connected to death: "Cosmic time is ... time which is torn apart into present, past and future.... [Such time] is diseased, and it does an injury to human existence. Death is connected with the disease of time".[48] What Berdyaev means here by death is the death of the individual, of the

---

this overview I will concentrate on Berdyaev's later concept of time, as we find it in *Slavery and Freedom*.

44 Berdyaev, *Slavery and Freedom*, 6; for the following discussion, cf. 257–65.

45 "To speak of time, is not always to speak of one and the same thing. Time has a variety of meanings and it is needful to make distinctions. There are three times: cosmic time, historical time; and existential time, and every man lives in these three forms of time"; ibid. 257.

46 Ibid. 257–8.

47 Ibid. Berdyaev follows Aristotle in arguing that the "character of change is the origin of the character of [cosmic] time"; cf. Aristotle, *Physics*, IV, 219b and 223b. Berdyaev argues that Greek philosophy hardly extended beyond perceptions of this most elementary level of cosmic time: "The Greeks were primarily concerned with the apprehension of cosmic time; among them the aesthetic contemplation of the cosmos predominated, and they almost failed to apprehend historical time"; Berdyaev, *Slavery and Freedom*, 258.

48 Ibid.

personal. While circular, cosmic time is inherently the time of death and rebirth, of alternating change, this change and rebirth may hold promise for the species or race, but represents destruction and death for the individual. The particularity of the individual is disconnected from and inherently rejected by the unchanging kernels of mythic time in its eternal return.

> [While cosmic time] knows a periodic spring of the revival of life ... that revival takes place not for those whom death has carried away, but for others. Victory over death is impossible in cosmic time. The present, which cannot be seized because it falls between the past and the future, annihilates the past in order to be itself annihilated by the future. In cosmic time the realm of life is subject to death.... Cosmic time is death-dealing not for the race but for personality; it desires no knowledge of personality and takes no interest in its fate.[49]

In contrast to circular, cosmic time, historical time is a human creation[50] and linear in nature: "Historical time is symbolized not by the circle but by the straight line stretching out forwards. The special property of historical time is precisely this stretching out toward what is coming, this reaching forward to determine".[51] Thus historical time (though still subject to cosmic time and divided into years and centuries) represents for Berdyaev the core human "revolt" against the uncaring and destructive circularity and repetitiveness of cosmic time. Historical time desperately seeks to maintain novelty and the "individually particular". In it, the active subjectivity of human personhood reacts against the bounded circularity of cosmic time. Yet it is precisely this human striving which leads to the downfall of historical time: "personality is wounded and enslaved by historical time in a new way, and at times it even looks to a transition to the cosmic sphere of existence for deliverance from the captivity of history".[52] The problem, as Berdyaev identifies it, is that historical

---

49 Ibid. 258–9.

50 "Historical time has a closer connection with human activity than cosmic time"; ibid. 259.

51 Ibid.

52 Ibid. This underlying insight is shared by Eliade, who presents "religious man" as escaping out of the linear "historical present" back into "circular time", which

time is ultimately defined precisely by the subjectivity of human perception, by Augustine's *memoria* and *expectatio*.[53] Rather than being neutral, these two dispositions polarize historical time so that "historical time is at once conservative and revolutionary" and torn between these two poles. Because of this deeply subjective (and sociopolitical) link, "historical time gives birth to illusions; the search in the past for what is better, truer, more beautiful, more perfect (the illusion of the conservative) or the search in the future for the fulness of achievement and the perfection of meaning (the illusion of progress)".[54] The subjective perception of an idealized, remembered past or idealized expected future prevents any significant existence in the present: "Historical time is time which is torn to pieces. It does not find completeness in any kind of present.... In the present man does not feel the fulness of time, and he seeks it in the past or in the future".[55] The present becomes vague and diffuse, importing into it the past and future (as we saw in Müller's matrix) while simultaneously siphoning off the concrete significance of the present. This is the seductive illusion of history that can only be defeated by that which Berdyaev calls "existential time".

Whereas cosmic time is represented by the circle and historical time by the straight line, Berdyaev's third time, "existential time", can be represented by a point[56] or by a vertical line that cuts perpendicularly across historical time[57]. What Berdyaev seeks to express with this image is the way in which existential time is an *interruption* of time, a "*Kairos* ... the

---

for Eliade represents the realm of cultic security: "reversible and recoverable, a sort of eternal mythical present"; cf. Eliade, *The Sacred and the Profane*, 69–70.

53 "Historical time strains forward in determination toward what is coming ... [and] is also connected with the past and with tradition which establishes a link between periods of time. Without that memory and that tradition in the inner sense of the word there is no history"; Berdyaev, *Slavery and Freedom*, 259.

54 Ibid. 260.

55 Ibid.

56 "Existential time may be best symbolized not by the circle nor by the line but by the point. That is precisely what is meant by saying that existential time can least of all be symbolized by extension. This is inward [time], not exteriorized in extension, not objectivized"; ibid.

57 "Everything which is accomplished in existential time, is accomplished along a vertical line, not a horizontal"; ibid. 261.

irruption of eternity into time, an interruption in cosmic and historical time".[58] Whereas Barth's eternity touches only tangentially upon temporality, and Pannenberg's eternity is the omnitemporal collection of all times, here Berdyaev sees eternity as the disruptive *fulfilment* of time. Yet this form of time should not be understood as objective (like cosmic time), but seen in a radically subjective and "personalist" way, located in the subjective perception of the individual. Existential time "is the time of the world of subjectivity, not objectivity. It is not computed mathematically, it is not summed up nor divided into parts. The infinity of existential time is a qualitative infinity, not a quantitative".[59]

This brief overview of subjective models of time shows us the amazing power that they hold, a power that comes to fulfilment in Berdyaev's system. In their focus on the human subject, these subjective models can do justice to our deep sense of radical integration into the flow of time. We are historical beings existing in and shaping the flow of time. Yet we not only shape and determine our experience of time but are also primarily shaped by and even trapped within the confines and pressures of temporality, in search for the cultic or existential moment that will break us out of the cycle of cosmic time or the uncaring progress of historical time. What subjective models do best is characterize our personal human experience of existence in the buffeting flow of time.

Yet here we also begin to see the difficulties that subjective concepts of time leave us with, a hurdle that has prevented subjective models from gaining traction in the broader scientific discourse as well as within theology itself. The first danger with temporal concepts based on the subjectivity of personal (and even communal) experience is that they impact on our belief in a shared reality and, when taken to extremes, set us on the road towards a fragmenting form of social constructivism (in its communal form) or solipsism (in its individualized or personalist form). If our human goal is to gather together in truth-seeking communities then there must be some perception that the truth which is sought is at least in some respects independent of human manipulation. Second, and closely related, we would want to ensure that our subjective models are not (at their worst) the result of individual or communal self-delusion.

58 Ibid. 260.

59 Ibid.

While the human mind and subjective consciousness are incredibly powerful forces that help us to conceptualize and understand the world, we cannot simply assume that such subjectivity is automatically good, beneficial, or supportive. Our subjective experiences both of time and reality can be skewed or distorted, infected by self-delusion and self-deception. There is no guarantee that these distorted perceptions of "existential time", to use Berdyaev's term, are positive, or will necessarily elevate and exalt us into some divine and eternal sphere. The fallenness of the human mind and human subjectivity can sink us into traumatizing and destructive experiences which disconnect us from beneficial contact with others, or distance us from reality. "Joy and triumphant rapture" may well lead to a subjective experience of the eternal, yet it is precisely this tendency that needs to be carefully evaluated, since "every state of ecstasy", even delusional, self-abusive and destructive ones, may lead "out from the computation of objectivized mathematical time and ... into existential qualitative infinity".[60]

Third, while Berdyaev agrees that it "would be untrue to say that existential time is identical with eternity" and that it is merely "an emergence into eternity",[61] he still sees this emergence as a positive step, one which is radically individualized in each person's "own moments of eternity". Yet the protraction of existential time has nothing in common with the protraction of objectivized time, cosmic or historical. This protraction depends instead upon *the intensity of experience* within present human existence.[62] This final retreat back into individual or communal consciousness represents a major problem for a Christian tradition that sees God as the Lord of the living and the dead. When we come to issues of soteriology and eschatology, God's eternity and the fulfilment and perfection of time and creation are not to be limited to a subjective experience in an individual's subjective consciousness but, in the objective creation of new heavens and new earth, a transformative fulfilment not only for all

60 Berdyaev, *Slavery and Freedom*, 261.

61 Ibid. 260, 259.

62 Ibid. 261. In this respect, Pannenberg is correct to see the problem facing personalism as a retreat into the inner self, arising as a reaction against the objectification of modern life and thought, Wolfhart Pannenberg, "Person", in K. Galling et al. (eds.), *Religion in Geschichte und Gegenwart* (Tübingen: JCB Mohr/Siebeck, 1958), v. 230–5 at 234.

non-conscious creation, not only for all living human beings, but also for all who have died, not just those who are able in this life to achieve a personal, subjective experience of transcendence. For the Christian traditions which aim at such fulfilment and redemption for *all* creation, shallow and temporary experiences of rapture and fulfilment in the subjective time of individual consciousness are not enough.

In the end, Berdyaev's example points us to a deep theological discomfort with all subjective models of time due to the suspicion that they not only disconnect us from shared realities but leave us with an escapist agenda for a personalized self-salvation that raptures the living out of the suffering of time. The Christian hope of salvation, resurrection and recreation addresses the pain of those who have suffered in time and have been destroyed by time—those beyond the boundary of a personalist, subjective disposition. The Christian eschatological hope in new creation directs us to an external, objective act of God, overcoming the alienation and destruction of time. Replacing such a transformative eschatology with personal and subjective experiences of rapture transforms Christianity into an opiate for both the individual and the masses.

Yet can the objective models of time that physics and Pannenberg offer us lead us to anything better? Let us turn now to a detailed examination of the temporal theories put forward by both special relativity and Pannenberg, and the mutual conflicts they both face within the academic discourse on time.

CHAPTER THREE

ENGAGING SPACE-TIME

3.1 Philosophical Impetus for the Development of Space-Time

When we try in a simple way to conceptualize our personal experiences of time, we might choose one of two popular approaches. The first—familiar to us from the segmented world of Cartesian graphs—dissects the world into three-dimensional spatial segments, each of which corresponds to *one* present moment. These column segments of the world ("world-wide instants") are then stacked together in a line from left to right, past to future, with each empty column of time waiting for its own particular contents to arrive. Taken together, the collection of these instants is seen as forming together the expansion of time, the empty and waiting horizontal axis of temporality that stretches out before us. We then imagine ourselves experiencing time by moving along the length of this axis, moving from one temporal column (from one "present" moment) into the next. This view offers us a picture of a dynamic self moving within and across the stationary axis of time, a Newtonian model where objects move within the absolute and empty frame of time. Here time is objective, existing beyond us, and in which we move and exist. Time here does not really *pass* or move, rather it is we and everything else that move within this stationary framework. As Kant argued: "Time passes not, but in it passes the existence of the changeable". Time, therefore, "is itself unchangeable and permanent".[1]

The second image, hints of which we see reflected in Augustine's subjective view of time, essentially reverses this picture and instead imagines the consciousness of the self as the stationary and unmoving rock in the

1 Immanuel Kant, *Critique of Pure Reason,* trans. J. M. D. Meiklejohn (1781; London and New York: Everyman, 1979), 120.

centre of the flowing stream of time. The individual is always stationary in the present, is always existing in the present. Past present, present present, future present flow towards this stationary point of existence.

Both views seek to achieve two goals. First, they try to focus upon and do justice to our basic intuition into the movement, change, and progress of the world around us, and use the concept of time as the measure for that movement. Second, they reflect a basic belief in the fundamental separation between temporal and spatial dimensions, between the constant and stable physicality of objects and our perception of objects in motion. This separation seems in line with human consciousness and our basic sensory experiences: "When I close my eyes and retreat into my inner mind, I feel myself enduring, I do not feel myself extensive. ... [Space] is always appreciated as something external".[2] Space and time appear to us as separate dimensions. In the first model of time, time itself takes primacy over space (since the world appears to progress along the empty axis of time). "In ordinary human experience, space is the order of togetherness of simultaneous phenomena, and time is the order of their sequence. It has often been said, with good reason, that time is the more fundamental of the two, because the concept of space as order of togetherness presupposes already the temporal notion of simultaneity".[3]

---

2 Eddington, *The Nature of the Physical World,* 60.

3 Pannenberg, "Eternity, Time, and Space", 98. "The differentiation of space from time is also a work of reflection that distinguishes the togetherness of things in space [i.e. simultaneity] from their following one another in time [i.e. consecutiveness]. The concept of time proves to be basic in this regard, for it is constitutive for that of space. The simultaneity of what is different constitutes space"; Pannenberg, *Systematic Theology,* ii. 90. As a simple reflection on "ordinary human experience", this represents a typical subjective concept of time, unsurprising if we follow it back to its roots in Droysen's statement that "Nicht objektiv scheiden sich die die Erscheinungen nach Raum und Zeit: unsere Auffassungen unterscheidet sie so, je nachdem die Erscheinungen sich mehr dem Raum, mehr der Zeit nach zu verhalten scheinen. Bestimmtheit und Inhalt gewinnen die Begriffe Raum und Zeit in dem Maße, als das Nebeneinander des Seienden, das Nacheinander des Gewordenen wahrgenommen, erkannt erforscht wird"; Droysen, §1 *Grundriß* (1882 ed.): J. G. Droysen, *Historik: Rekonstruktion der ersten vollständigen Fassung der Vorlesungen (1857), Grundriß der Historik in der ersten handschriftlichen (1857/58) und in der letzten gedruckten Fassung*

This divided approach towards space and time struck upon serious complications in 1904 with Henri Poincaré's *Principle of Relativity,* a concept that was then later developed by Einstein and Minkowski, especially between 1905 and 1908.[4] Poincaré's work on the relations of the laws of physics as they apply to the so-called "observer at rest" and "observer travelling with uniform motion" led to the conclusion that one cannot have any way of knowing which of the observers is actually at rest and which is moving. This conclusion represented a first step in the movement away from Newton's understanding of space as an absolute three-dimensional container separated from time, and towards the realization of the interconnectedness of all the quantities of physics. Einstein's development of this work led to a new framework of reference, resulting in the obsolescence of Newton's absolute time and space as ontologically separate realities. Time was now no more absolute than mass or velocity, nor could it be fundamentally differentiated from space. On the contrary, relativity argued that time was intimately combined with space; characteristics such as time, length or mass, which were believed to be physical constants, were now found to be dependent upon one's own frame of reference. Thus in the early days of this revolution, physics was caught up in a dual movement: it had abandoned the Newtonian separation of time and space, yet lost with it a sense for the absolute, stable quantities of the physical reality around us.

In 1908, amid this confusion about the inherent relativity of the universe, the Göttingen mathematician Hermann Minkowski, in a research paper entitled *Raum und Zeit*[5], outlined his method for finally regain-

(*1882*), ed. by Peter Leyh (Stuttgart: Fromann-Holzboog, 1977), 421.

4 I am not going to enter here into the long debate on whether the theory of relativity should be credited properly to Poincaré or Einstein. Poincaré's work concentrated on the problematic attempts to establish the velocity of the earth relative to the ether, and as Dirac notes, "Poincaré looked upon his work as flowing from experiments and, like Lorentz, did not understand that a fundamental new physical principle was involved"; P. A. M. Dirac, "Einstein and the Development of Physics", in C. M. Kinnon, A. N. Kholodilin and J. G. Richardson (eds.), *The Impact of Modern Scientific Ideas on Society: In Commemoration of Einstein* (Dordrecht: Reidel, 1981), 13–23 at 15.

5 H. Minkowski, *Raum und Zeit: Vortrag, gehalten auf der 80. Versammlung Deutscher Naturforscher und Ärzte zu Cöln am 21. September 1908* (Leipzig; Berlin:

ing the absolute: the geometrical unification of spatial and temporal dimensions in a four-dimensional space or "manifold".[6] Whereas Einstein's work had fragmented the universe by displaying the independent and relative aspects of reality, Minkowski reunified them by showing that they were simply *different frames of reference to the same absolute object.* In his famous introduction to the 1908 paper, Minkowski declared: "The views of space and time which I wish to lay before you have sprung from the soil of experimental physics, and therein lies their strength. They are radical. Henceforth, space by itself, and time by itself, are doomed to fade away into mere shadows, and only a kind of union of the two will preserve an independent reality".[7] Minkowski's dramatic prediction would soon be realized.

According to Minkowski's model, the union of space and time pointed to the existence of the universe as a solid four-dimensional reality which we then measure from differing points of reference. If we imagine a four dimensional cylinder, then a single frame of reference may slice across this solid at one "angle", producing a circular plane. But other frames of reference cutting across the cylinder at various other "angles" would produce a variety of different oval-shaped planes, the measurements in each differing from each other. Importantly, it is not that this variance in measurement points to different objects being measured, or different realities—they are simply variant "frames of reference" that all still deal with the same invariant, four-dimensional object. We are investigating and measuring the same invariable object, but those measure-

---

Teubner, 1909); reprinted in H. A. Lorentz, A. Einstein, and H. Minkowski, *Das Relativitätsprinzip: Eine Sammlung von Abhandlungen mit einem Beitrag von H. Weyl und Anmerkungen von A. Sommerfeld* (Berlin: B.G. Teubner, 1922), 54–66 = H. Minkowski, "Space and Time", in H. A. Lorentz, A. Einstein, H. Minkowski and H. Weyl, *The Principle of Relativity: A Collection of Original Memoirs on the Special and General Theory of Relativity,* trans. W. Perrett and G. B. Jeffrey (London: Methuen, 1923), 73–91.

6 "The appearance of a four-dimensional world in this subject is due to Minkowski. Einstein showed the relativity of the familiar quantities of physics; Minkowski showed how to recover the absolute by going back to their four-dimensional origin and searching more deeply"; Eddington, *The Nature of the Physical World,* 62.

7 Minkowski, "Space and Time", 75.

ments will change depending on our frame of reference. This combined space-time is unified and objective, but its measurement is unavoidably multireferential.

True to his own expectations, Minkowski's work on the special theory of relativity has since become the standard interpretation of Einstein's work and has been influential over the last century not only upon so-called "manifold theories" of time, but on divergent fields across the academic spectrum.[8] However, adoption of the principles of Minkowski space-time leads to particular assumptions and difficulties in the philosophy and theology of time. So it will be helpful to take a closer look at the development of the philosophy of Minkowski space-time, the principles underlying that development, and the questions and problems it seeks to answer.

3.1.1 Four-Dimensional Extension and the Absolute

If we begin with the familiar three-dimensional space of Cartesian geometry, we see how it imagines an empty space within which objects can be located along x, y and z axes.[9] While this is adequate for spatial dimensions, a fourth axis, t, would still be needed to signify the expanse of time. In a very basic way, simply adding such a fourth axis would allow us to locate events then within space and time by attributing to each event four co-ordinates: three for the spatial axes (x, y, z) and an additional co-ordinate t for time. Points along these axes would then correspond to differing degrees and differing kinds of remoteness from the origin or zero point on our Cartesian graph: positions along the x-axis, for example, would correspond to an x-measure of *spatial* remoteness either to my left or right (positions "elsewhere", also the case for the dimensions of the y- and z-axes); positions on the t-axis would correspond to *temporally* remote events (positions "future" or "past"). Thus, in a Cartesian

8 For a general overview into the social and academic impact of space-time relativity, see Kinnon, Kholodilin and Richardson (eds.), *The Impact of Modern Scientific Ideas on Society.*

9 The following discussion draws on the now somewhat dated, but historically important description of space-time and relativity offered by Eddington in his Gifford Lectures, *The Nature of the Physical World.*

model, we could attribute a four-fold set of co-ordinates to each object that we encounter: *left–right, above–below, in front–behind,* and *past–future.* However, these co-ordinates of remoteness can only be meaningful in relation to a frame of reference. In the Cartesian framework, I can set myself at zero—the point where all the axes meet—and describe this point as *here.* If I use this method for locating events around me, then the Cartesian co-ordinates provide a good representation of an object's degrees of remoteness from myself: 0 + 3 metres to my right, 0 – 2 metres behind me and 0 + 1 metre above. In a simple way, we might then try to imagine these three-dimensional planes running in sequence to create a crude fourth dimension to represent time. If we do this, the Cartesian zero point of reference ("myself") would stretch out in time to become a "myself *line*", extending temporally through the world from past to future.

The Cartesian *point* of reference still exists within this temporally extended line (as the discrete centre of each three-dimensional spatial plane). The current point in the current three-dimensional plane would be "Here-Now". A point on the line below "Here–Now" would represent "Here–Yesterday", and "Here–Tomorrow" would lie above the midpoint. If I look around from my "Here-Now" I see a range of other events which I believe are also occurring now, so I draw a line across from *my* "now" through these events to include them, to signify that these events while not simultaneously occurring *here*-now, are occurring *elsewhere*-now, within the same three-dimensional plane of space which I am occupying now.

This simple method for locating events in four-dimensions ran into serious problems in 1675, when the Danish astronomer Ole Rømer discovered that light does not move instantaneously through the universe but takes time to travel.[10] Rømer's discovery revealed our "instantaneous now" to be a fiction: what we *see* now could no longer be taken as *occurring* now, but must be understood as a delayed event. We must distinguish between two separate events, that is the occurrence of the original event in the "external world", and the occurrence of our perception of

10 Rømer discovered the finite speed of light in his examination of Jupiter's moons. For a helpful retelling of Rømer's discovery, see John Tyndall, *Light and Electricity: Notes of Two Courses of Lectures Before the Royal Institution of Great Britain* (New York: D. Appleton, 1871), 19–20.

that event "here–now". "The second event was in our bodies Here–Now; the first event was neither Here nor Now."[11] In Kantian terms, the discovery of the speed of light forced a separation between the "noumenal event" and the "phenomenal event"; in historiographical terms, with Rømer the natural sciences found themselves with a distinction between *Historie* and *Geschichte,* between "actual" event and "interpreted" event.

To compensate for the effects of Rømer's discovery on the process of locating objects in time, we could try to amend the Cartesian diagram by adding "Seen–Now" lines which would angle back towards the past from our horizontal, assumed "now line" (the angle of the line matching the delay attributed to the speed of light). Yet these "Seen–Now" lines are no more dependable than the previous "now" line, since travellers moving with differing velocities cannot agree on what they see now (that is, what is happening *now*): what for one person is a past event will for the other still be a simultaneously occurring event, and for another still a future event. Our instantaneous *presents* or *world-wide instants* are "not natural cleavage planes within time; there is nothing equivalent to them in the absolute structure of the world".[12] We tend to think of the world as stratified into successive instantaneous moments each with its own absolute, universal *now.* But it is this stratification into a multitude of three-dimensional Cartesian planes extending in four dimensions—the misguided impression "of the enduring world as a three-dimensional space leaping from instant to instant through time"— that the manifold of Minkowski space-time, with its solid, four-dimensional continuum, rejects.[13]

In developing his geometry of the space-time manifold, Minkowski removes the "now line" since there is no absolute present, only a range of relative "presents" for each respective observer.[14] Together with the

11 Eddington, *The Nature of the Physical World,* 53.

12 Ibid. 56.

13 Ibid. 47.

14 I have avoided speaking of "absolute simultaneity" here since there is a great deal of confusion regarding the possibility of simultaneous events in special relativity. Simultaneity is relative to a frame of reference. Events can be simultaneous only relative to a single moving frame of reference but are not simultaneous with respect to a second frame. It is not that special relativity disallows the simultaneity of events, rather it stresses that there can be no simultaneity across differing frames. Two points are vital here. First, if by "the present" we

"now" line, Minkowski also removes the "myself line" representing the subject, since neither the subject nor the present are inherent or essential to the world. What remains are the Seen–Now lines, which correspond to the limits represented by the speed of light. Since all inertial observers from Here–Now will agree on what is "seen now", these lines are taken as absolute. If one imagines this diagram in four dimensions, then the speed of light (represented by the "seen now" lines) form an unbreakable cone extending from the past and having its apex at Here–Now. These lines do not stop here-now, but extend into the future connecting the cone of the past with the inverted cone of the future. These four-dimensional twin cones signify the absolute past and absolute future, with the point of each cone touching at Here-Now. Since nothing can break out of these cones (to do so they would have to travel faster than light), that which lies outside these limiting cones (in the "neutral wedge") can be labelled "absolute elsewhere".[15] These places are neither Here–Now, nor fall within the boundaries of the cones of the past and future associated with this particular Here–Now.

3.2 Theological Impact and Importance of Space-Time

This brief overview of the development of Minkowski space-time allows us to grasp the origins of three important issues which a theology of time will need to address in its discussion with the sciences: restriction, invariance, and determinateness. Yet we will also see that these issues already have a presence in theology, having entered through the apocalyptic traditions. Indeed, we find that biblical apocalypticism displays each

mean the simultaneity of two or more events, then it is clear that special relativity leads the present to become increasingly slippery if we try to enforce this present across various frames. However, second, while discussions of simultaneity in physics are often concentrated upon *point-like* events, as soon as we introduce extended, durational events then it is no longer invalid to speak of simultaneity across frames, so long as the duration of the event is sufficiently large. For example, at the greatest possible spatial separation on earth, any event extending beyond approximately 0.1 seconds will meet the requirements for simultaneity; cf. Eddington, *The Nature of the Physical World*, 58.

15 Ibid. 57.

of these aspects that are so vital in Minkowski space-time, making the apocalyptic an important point of contact between theology and physics in this discussion.

3.2.1 Restriction

Minkowski space-time unifies the three spatial dimensions with the temporal dimension to produce a four-dimensional manifold. In this manifold, the motion of a particle is restricted. Since nothing can travel faster than the speed of light, it is not possible for an object to move from the temporal cones of past or future into the neutral wedge of "absolute elsewhere". If a normally-charged particle attempts to reach a spatially distant position, then regardless of how fast it moves, it can only ever move *forwards* from its current here–now into the upper cone of the absolute future. Even if the particle were able to move at the speed of light, it still would not be fast enough to break out of the cone of the absolute future into the neutral zone of "absolute elsewhere". Since a particle cannot travel faster than the speed of light, it cannot break out of the limiting cones of its own spatiality and temporality (based on its own initial location) to reach an event located spatially and temporally in "absolute elsewhere".

Three important consequences arise here. The first touches on the theological hope in a divinely created, open and radically novel future. The absolute future of Minkowski space-time re-grounds the radical finitude both of creaturely existence and the future. Rather than an open future of infinite possibility (and divine power), the future represented by Minkowski space-time is limited: it is limited by the past and present, or more accurately by an object's current extension within the manifold. The absolute future is limited in a very real way due to an object's spatio-temporal status. In human terms, one's future and spatial location within that future are strictly limited by one's particular location here–now and determined by one's spatio-temporal extension in the four-dimensional manifold. The past and present provide the grounding for the unavoidable limitation of the future. Due to our nature as finite physical entities, the future that we experience is in no way radically open and contingent, but is rather always limited in line with the initial conditions of our existence Here-Now. On the one hand, this limitation is not strictly *qualitative* (affecting the nature of the event), but only *locative* (determining the spatial and temporal *location* of that event). On the other hand, this lends communal stability to the future, since all objects sharing the same (broad) frame of reference[16] will exhibit a common set of future possibilities. This shared future stems directly from (and is determined by) our common past within this broad frame of reference. Thus movement is not only restricted—having a fixed direction of movement from the absolute past into the absolute future[17]—but that movement can be standardized, offering a high degree of predictability.

The second consequence impacts upon the theological language of divine omnipresence. We noted above that nothing can move laterally, to break out of the temporal cones into the neutral zone of "Absolute Elsewhere" since to do so would require that it travel faster than the speed of light. This leads us to the problem of "simultaneous signalling".

16 For our earth-bound purposes, i.e. events across a temporal wedge of approximately 0.04 seconds (dividing the equatorial diameter of the earth by the speed of light).

17 The limits of the cones of light may seem to disallow any concept of time which is not directly linear or having the arrow of time pointing into the future. However there are certain provisos. First, the spherical nature of a universe pictured as being "finite yet unbounded" could lead to a circularity in time similar to the circularity of space. Eddington acknowledges that "If we go a very long way to point *a* in one direction through the universe and a very long way to point *b* in the opposite direction, it is believed that between *a* and *b* there exists a linkage of the kind indicated by a very small code number; in other words these point reached by travelling vast distances in opposite directions would be found experimentally to be close together". Yet he concludes that while the world is closed in its spatial dimensions, it must be open at both ends with regard to time: "There is a bending round by which East ultimately becomes West, but no bending by which Before ultimately becomes After"; Eddington, *The Nature of the Physical World,* 88–9. Second, Richard Feynman has suggested that particles moving with negative energy can be said to have world lines which move backwards through time, from the cone of absolute future into absolute past; R. P. Feynman, "The Theory of Positrons", *Physical Review,* 76, 749–59 (1949), reprinted in R. P. Feynman, *Quantum Electrodynamics* (Boulder, CO: Westview, 1998), 167–77, esp. 168.

> The assertion made by relativity theory is [that]: Neither matter, nor energy, nor anything capable of being used as a signal can travel faster than 299,796 kilometres a second [i.e. the speed of light], provided that the velocity is referred to [its appropriate frame]. ... We are bound to insist on this limitation of the speed of signalling. It has the effect that it is only possible to signal into the Absolute Future. The consequences of being able to transmit messages concerning events Here–Now into the neutral wedge are too bizarre to contemplate. Either the part of the neutral wedge that can be reached by the signals must be restricted in a way which violates the principle of relativity; or it will be possible to arrange for a confederate to receive the messages which we shall send him to-morrow, and to retransmit them to us so that we receive them to-day.[18]

Thus we saw that the speed of light provides the boundary which locks an event within the limits of its own frame of reference. But is this limitation one that affects God, and is it sensible to speak of God "existing" within and limited to a particular frame of reference as with other substantial entities?

John Polkinghorne and Wolfhart Pannenberg both hint at the possibility of an exemption for God: that God would be freed from the limitation imposed by the speed of light (and thus not affected by the problem facing simultaneous signalling) since God already exists across *all* spatio-temporal frames of reference rather than merely in one. The suggestion here hangs on the concept of God's omnipresence, that is that God is not *localized* (point-like) within a frame of reference but rather expansively "omnipresent within it". This would thus allow God to bridge frames, be free of the limitations of time, and achieve "simultaneous signalling".[19] Pannenberg pursues this idea, arguing that "God's eternity ... is itself simultaneous with all events in the strict sense. God does not need light to know things. Being omnipresent, he is with every creature as its own place".[20] God's omnipresence overcomes the distinction between "now" and "seen–now"—which we were forced to accept due to the speed of light—nor is an omnipresent God required to *move* from one frame to another, since God is already *eternally and simulta-*

18 Eddington, *The Nature of the Physical World,* 65.

19 Polkinghorne, *Belief in God in an Age of Science,* 71.

20 Pannenberg, *Systematic Theology,* ii. 91.

neously present in all spatial and temporal frames.[21] Pannenberg's appeal here to the omnipresence of God in order to secure both the notion of simultaneity and of God's eternity as "the undivided present of life in its totality"[22] foreshadows his development of the idea of eternity (as the omnitemporality of all times) as God's frame of reference, and the extensiveness of the Spirit in a cosmic field of force[23]. While this move is certainly elegant, it remains questionable whether this suggestion subsumes the nature of God too much into the physical nature of creation, leading us to the problems of pantheism or an ontotheology which fails sufficiently to distinguish the nature of God from that of God's creation. In this respect, we gain no great advantage over a pseudo-Newtonian concept of God as force.[24] Furthermore, special relativity's integration of space, time, and matter presents difficulties for concepts of God that want to see God either as *immaterial yet temporal* (that is, not localizable in space but existing in time!) or *material yet a-temporal* (that is, localizable in space but not limited temporally!) since space and time are now conceived of as inseparable.

Interestingly, the radical integrity of Minkowski space-time (that is, its unification of space, time, and matter into a single four-dimensional manifold) reflects aspects of the biblical apocalyptic traditions. One of the marking characteristics of that tradition was its unification of all time, history, and reality into a single totality that spanned from the beginning of creation to its end.[25] Space-time shares with the apocalyptic the pos-

21 "[F]or God eternity is the simultaneous presence and possession of the wholeness"; Pannenberg, "Eternity, Time, and Space", 97.

22 Pannenberg, *Systematic Theology,* ii. 92.

23 We will discuss these points in the following two chapters.

24 The "idea of a god as *a force* pushing and pulling matter about alongside the other forces of nature is decidedly unappealing to scientists—and to many theologians, since it reduces God to an aspect of nature in competition with other aspects"; Paul Davies, "Teleology without Teleology: Purpose through Emergent Complexity", in Philip Clayton and Arthur Peacocke (eds.), *In Whom We Live And Move And Have Our Being: Panentheistic Reflections on God's Presence in a Scientific World* (Grand Rapids: Eerdmans, 2004), 95–108 at 97.

25 Wolfhart Pannenberg, "Redemptive Event and History", in id., *Basic Questions in Theology,* i (London: SCM, 1970), 15–80, esp. 20–1; see also Wolfhart Pannenberg, "The Biblical Understanding of Reality", in id., *Faith and Reality,* trans.

tulate of a block universe which completely encompasses the universe's component parts.

3.2.2 Invariance

The invariant nature of Minkowski space-time is a consequence of its four-dimensional nature, or the limitation of the space-time manifold to only four dimensions. When we conceive of *time* as moving or flowing, then this requires that time itself must have something (such as an additional fifth dimension) in which to flow. This is the case if we continue to imagine the universe as a three-dimensional space leaping from instant to instant through time: three-dimensional segments of the world "flow" inside the box-like framework provided by "absolute time". However, Minkowski's unification of space and time into a single absolute prohibits such a view. Instead of three-dimensional objects existing in the container of time, the Minkowski universe presents a collection of *invariant four-dimensional objects* which together form an invariant *block universe:* a four-dimensional extension that remains constant, already existent, and which cannot "change": "four-dimensional space-time does not change. Something can only change *in time.* But space-time is not 'in time'. *All* of time is [already] there ... *past, present, and future.* It's static."[26]

As such, the invariant nature of Minkowski space-time demands that objects be considered as *complete* four-dimensional entities, that is, not as entities temporally limited to the present moment or as changing through time. To think only of an object in its present moment is shown then to be an unhelpful reductionism, an abstraction which severs that object's here–now existence from its past and future temporal existence. At the ontological level, this is one of the most powerful advantages to the Minkowski model since it shifts epistemology and ontology beyond any notion of object-entities suddenly popping in and out of existence with the flow of time, as if object *a* is somehow ontically different to-

John Maxwell (London: Search, 1977), 8–19 at 15.

26 Russell Stannard, "God in and beyond Space and Time", in Philip Clayton and Arthur Peacocke (eds.), *In Whom We Live And Move And Have Our Being: Panentheistic Reflections on God's Presence in a Scientific World* (Grand Rapids: Eerdmans, 2004), 109–20 at 113.

day from what it was yesterday, and will be again something different tomorrow, namely a "future" object somehow "yet to exist". Matter in the Minkowski universe does not simply appear and disappear, but has a radical permanence.[27] The future is neither an empty space waiting to be filled, nor a vacant tense waiting to be populated by an advancing present. The Minkowski universe does not present or allow such stratified objects which dissolve with the past and are re-formed with each present moment. The Minkowski universe is not created "new every morning". Rather it presents stable objects which exist and extend as a *four-dimensional whole,* and thus constitute four-dimensional space-time. To think of an object only in its *present* state without including its entire existence as a four-dimensional solid extending from its absolute past into its absolute future is therefore an abstraction which fails to do justice to the temporal extension of that object.[28]

This consequence of the space-time manifold represents one of the most significant steps away from Newtonian cosmology. Rather than postulating an open space in which objects *grow* and *develop, change* and *advance,* the Minkowski space-time manifold is a *static entity* consisting of *invariant four-dimensional solids.* Expressed in theological terms, the act of creation is not then an establishment of first principles for objects which may then run their own course in line with the classical laws of nature. Rather creation for the Minkowski universe means *the coming into*

27 It should be stressed again that such a concept of the block universe was foreign to Einstein's early work. Einstein's famous 1905 papers established the special theory of relativity, but it was Minkowski who finally developed Einstein's work into the concept of the block universe "as a philosophical consequence of relativity. This step depended on Minkowski's formulation of the concept of space-time". Only "[a]fter some initial hesitation, Einstein eventually accepted the fruitfulness of this notion"; Friedel Weinert, *The Scientist as Philosopher: Philosophical Consequences of Great Scientific Discoveries* (Berlin: Springer, 2005), 128.

28 "We [must] rid our minds of this illusion of stratification. The world that then stands revealed, though strangely unfamiliar, is actually much simpler. There is a difference between simplicity and familiarity. A pig may be most familiar to us in the form of rashers, but the unstratified pig is a simpler object to the biologist who wishes to understand how the animal functions"; Eddington, *The Nature of the Physical World,* 56.

being of these four-dimensional solids in their entirety, en bloc. The space-time manifold is thus a static universe, not in the sense that there is no relative movement, but in that no *variance* can be introduced into these four-dimensional solids once they exist. Thus in the Minkowski universe there is no proper advancement through time; the "flow" of time is revealed as an illusion.

This focus on invariance reveals a second close affinity between Minkowski space-time and the apocalyptic world view . We have seen that apocalypticism unifies history and reality into a single totality spanning from creation to its end. Yet it is precisely that end of the totality which occupies the apocalyptic perspective. The totality *cannot* change and, indeed, for the apocalypticist there is no hope or expectation that that totality *will* change.[29] The apocalyptic focuses rather on the processes of destruction of that totality, and God's promised rebirth of that totality in *new* heavens and *new* earth (cf. Isa 65:17, 66:22; Rev 21:1, 5). Likewise, the Minkowski universe is unable to accommodate the radical change of *invariant* four-dimensional solids which compose the four-dimensional manifold of space-time. But this deficiency can be addressed by hypothesized "successive cycles of *an oscillating universe,* expanding and contracting, with a Big Crunch before each Big Bang"[30] which provides for the total destruction of the current universe and its replacement with its own new heavens and new earth.

3.2.3 Determinateness

A consequence that arises from the concept of invariant four-dimensional solids is their inherent determinateness. If an object already exists with temporal extension, with its reality already extended from "past" into "future", then that future cannot be subject to change. In our Cartesian diagrams we can plot the position of an object over time, and interact

29 R. H. Charles, "Apocalyptic Literature", in J. Hastings et al. (eds.), *A Dictionary of the Bible Dealing with its Language, Literature and Contents Including the Biblical Theology* (Edinburgh: T & T Clark, 1951), 109–10.

30 Ian G. Barbour, "God's Power: A Process View", in John Polkinghorne (ed.), *The Work of Love: Creation as Kenosis* (Grand Rapids: Eerdmans, 2001), 1–20 at 16.

with that object to change its location in space. The motion of these objects testifies to the influence of force in accordance with the Newtonian laws of motion and the conservation of energy, the preservation of a body in its state of rest until it is compelled to change its location due to the external force I impress upon it. However, within the space-time manifold such a view represents our failure to think consistently in four-dimensions, and instead reverting to the reductionistic abstraction of the object existing at each "present" point. That the object has changed in space has less to do with my ability to interfere with the Newtonian course of objects within the empty containers of space and time. Rather, since the "future" already exists, I am as capable of changing the future state of the manifold as I am of changing its past. As an *invariant four-dimensional object,* the non-stratified object stretches from its absolute past into its absolute future. Its location in that "future" simply displays what its "future" "has always been", or more precisely: its path of extension through the space-time manifold.

This needs to be distinguished carefully from the system of rule by deterministic laws typical in classical, Newtonian physics, which sought to determine the movement of objects into a blank and empty future (and thus control that movement). In contrast, Minkowski's model presents extended and enduring four-dimensional objects as they already are. Whether their "future" states can be connected with their "former" states in any deterministic, law-like way remains unclear.[31] Minkowski's geometry describes the position of the extended object in space-time; but that the spatial reference should change when the temporal reference also changes is not a statement of causation.[32] The block-like nature of the space-time manifold presents non-stratified objects with a temporal extension that is already complete, and disconnects this from the Newtonian concept of inertial or influential force. This "modern understanding of time would seem to indicate that whatever the future holds, in some

31 "At this particular instant we might possess data regarding past events, but not the information that would allow us to predict with certainty what lies in the future"; Russell Stannard, "God in and beyond Space and Time", 114.

32 The key concept here is that of disclosure or, in theological terms, "revelation". The "future, in itself, is *not* uncertain", however, we "do not *know* what the future holds because all we are directly experiencing is what exists at this particular instant, now"; ibid. 113.

sense it is already there, fixed, waiting for us to come across it".[33] Theologically, this implies that Minkowski space-time disallows efficacious temporal intervention (human or divine) since objects are already existent as extended and invariant four-dimensional solids. From the perspective of the space-time manifold, the liberal Renaissance Man who actively shaped the world around him becomes as much a fiction as the interventionist God who acts through a universal *Heilsgeschichte.*

Again, we strike here upon the particular affinity to the apocalyptic world-view, which sees the course of the reality "totality" as determined from the beginning and allowing of no efficacious human intervention.[34] As with the space-time manifold, there is a determinateness which disallows change to that system as it was initially established. The future for apocalypticism (that time between the present and the divine institution of new heavens and earth) is not a realm of openness and contingency, but merely discloses that divine plan which was firmly established from the beginning of time.[35] Similarly for the Minkowski universe, the future does not represent radical contingency but merely the *disclosure* or *revelation* of what the state of the four-dimensional manifold already is and has always been.

3.3 Pannenberg's Critique of Space-Time

Despite the general dominance of special relativity in scientific discussions on time, the manifold theory of space-time and the concept of the block universe have not been without their critics.[36] A great deal of attention has been focused upon the static and determinate aspects of

33 Ibid. 114.

34 Charles, "Apocalyptic Literature", 110.

35 The apocalyptic looks to the "plan of God which is firmly established from the beginning of time and [is] gradually disclosed by history"; Jürgen Moltmann, *Theology of Hope: On the Ground and the Implications of a Christian Eschatology,* trans. J. W. Leitch (London: SCM, 2002), 122.

36 A theological overview is offered by Stannard, "God in and beyond Space and Time". Since this section deals with the extent and limits of Pannenberg's critique, detailed exposition of all lines of opposition in the current discourse will not be presented here.

the space-time manifold, since these aspects present the greatest challenges to our theological concepts of hope and our common perceptions of time and the physical world. These critiques vary. On the one hand, there have been attempts to postulate additional cosmic dimensions in order to avoid the static nature of the four-dimensional manifold (that is, to find a fifth dimension in which the four-dimensional manifold of Minkowski space-time can exist and "flow").[37] This is a desperate approach, since any such fifth dimension (or "hypertime") simply shifts the problem back successively until we need an infinity of hypertimes.[38] On the other hand, more philosophical approaches, such as Henri Bergson's work on free will or Whitehead's process theory, provide some impetus for countering the determinateness of Minkowski's universe.[39]

---

37 "All movement is in relation to a time. If we say that time moves, this must be with respect to another 'hypertime', and since this hypertime presumably moves, it can only do so with respect to another, and so on. Similarly, if I say that I move through time, that is to commit myself to an infinite series of meta-times. If I ask: 'at what rate do I move through time? Minutes per ... ?', the blank can only be filled with some temporal term—say, 'minute'. But this 'meta-minute' to which my motion is now being referred, can only make sense against some further 'minute', and so on ad infinitum"; Jeremy S. Begbie, *Theology, Music and Time* (Cambridge: Cambridge University Press, 2000), 65 n. 93, referring to and building upon work by D. C. Williams. This tendency can also be witnessed in Whitehead's process theory. Whitehead postulates a spatialized set of relations for events with the temporal duration of that relatedness maintaining the pattern of that event: "Endurance is the repetition of the pattern in successive events. Thus endurance requires a succession of durations, each exhibiting the pattern"; Whitehead, *Science and the Modern World,* 125. Thus time is not a form of extensiveness, but simply the succession of epochal durations; ibid. It might seem nonsensical to speak of time as the "succession of durations", since this would suggest a supra-temporal state in which succession and duration occur (precisely the condition Minkowski space-time seeks to avoid). Indeed, Whitehead seems to move in this direction arguing that "in some sense, time ... extends beyond the spatio-temporal continuum of nature"; ibid. 124.

38 Cf. the much-critiqued attempt to do just this by John William Dunne, *An Experiment With Time* (Black, 1927), esp. part iv.

39 Whitehead's process philosophy has gained more traction compared to Bergson's approach since it refused to introduce a Bergsonian epistemological separation between the truths of reason and the truths of intuition, but rather

Reflecting a subjective understanding of time, Bergson and his followers argued that there exists an inherent aspect of change in the universe which mathematics and physical theory simply cannot grasp: understanding true change and the true nature of time is beyond the limits of scientific reason, which is only open to the intuitive power of consciousness.[40] Bergson was not alone in this view but was rather voicing an opinion that had already developed in continental philosophy, and

---

tackled the block nature of space-time with a new approach to ontology itself. Whitehead was one of the earliest critics of relativity theory. He had already presented a significant text outlining the bulk of his ideas on this theory in his *The Principle of Relativity with Applications to Physical Science* (Cambridge: CUP, 1922), and despite the relative newness of the idea in physics and the lack of penetration the theory had had by that time into the broader academic environment, by 1925 (with the publication of his *Science and the Modern World*) Whitehead already had a significant reputation as a critic of relativity. The ground-work for the long-standing (and still effective) polemic between the so-called "philosophers of the manifold" and progress philosophy was laid in this period, though it would become progressively more ingrained after Whitehead's (poorly-received) Gifford lectures from 1928 (all the more bitter given the wide acclaim of Eddington's Gifford lectures on relativity from the year before); Alfred North Whitehead, *Process and Reality: An Essay in Cosmology* (corrected edn.; New York: Free Press, 1978). See T. E. Hosinski, *Stubborn Fact and Creative Advance: An Introduction to the Metaphysics of Alfred North Whitehead* (Lanham, MD: Rowman & Littlefield, 1993) for a helpful overview.

40 "Now ... science cannot deal with time and motion except on condition of first eliminating the essential and qualitative element—of time, duration, and of motion, mobility. ... Treatises in mechanics are careful to announce that they do not intend to define duration itself but only the equality of two durations. ... To announce that something will take place at the end of a time *t* is to declare that consciousness will note between now and then a number *t* of similarities of a certain kind. And we must not be led astray by the words 'between now and then', for the interval of duration exists only for us and on account of the interpretation of our conscious states. ... [T]he interval of duration itself cannot be taken into account by science"; Henri Bergson, *Time and Free Will: An Essay on the Immediate Data of Consciousness,* trans. F. L. Pogson (Mineola, NY: Dover, 2001), 115–16. Cf. also Wolfhart Pannenberg, "Atomism, Duration, Form: Difficulties with Process Philosophy", in id., *Metaphysics and the Idea of God* (Grand Rapids: Eerdmans, 1990), 113–29 at 114–16.

which reflected the stance taken by the German historicist school, namely that history itself is not to be understood in and through scientific rationality but can only be understood through the historian's artistic intuition or *Verstehen*.[41] This "process of intuition ... is a form of interpersonal perception that transcends the rational", it "is not achieved ... by reason at all".[42]

Similarly, Pannenberg's own approach, as we saw at the beginning of this chapter, was that while the "concepts of space and time are important in physics and geometry ... their definition is not the exclusive prerogative of those sciences", nor is it certain "that the measurement of spatial and temporal relations exhausts the concepts of space and time".[43] Here we have an excellent encapsulation of the two lines of attack in Pannenberg's critique. The first notes that the natural sciences do not represent the sole or final authority on issues of space and time, since the "description of space does not belong exclusively to geometry but is also and even primarily a matter of philosophy".[44] The second aspect of Pannenberg's argument rightly stresses the limited focus of the natural sciences on the *measurement* of space and time, rather than on coming to understand the *ontology* of space and time. The "space-time concept of relativity theory does not unsettle [the] basic philosophical analysis of space and time" nor does it "completely reconstitute our notions of space and time".[45] Instead, Pannenberg agrees with William Lane Craig that, "At best, scientific accounts describe our *measures* of time, but not time itself"; "Curved space-time is just a geometrical model".[46] Both these lines of argument are powerful in their own ways: the first tacitly asserts that the view of time and space provided by the natural sciences is fundamentally insufficient to account for the breadth of human insight into these aspects of reality—insights stemming both from human experience and from the broader academy of sciences (such as

41 Cf. Bebbington, *Patterns in History*, 92–116.

42 Ibid. 93.

43 Pannenberg, "Eternity, Time, and Space", 97–8.

44 Ibid. 98.

45 Ibid. 99.

46 William Lane Craig, *Time and Eternity: Exploring God's Relationship to Time* (Wheaton, IL: Crossway, 2001), 66, 178; quoted by Pannenberg, "Eternity, Time, and Space", 99.

the humanities). The strength of the second line of argument is that it reflects an epistemological awareness that has long since arisen within the natural sciences themselves, namely that the manner in which the "exact" sciences proceed revolves solely around the practice of the reductionism of measurement. The problem arises when one begins to assume a positivistic *unification* of those measurements with the object being measured.[47] As Polkinghorne also notes, "There is no logically inevitable way to proceed from epistemology to ontology, from what we can know about entities to what they are actually like".[48] While "the vast majority of working scientists, consciously or unconsciously ... opt for critical realism, which one could define as being the attempt to maximize the correlation between epistemological input and ontological belief",[49] this is not always enough to avoid errors (particularly errors of substitution between two entities with identical, or near identical, values).[50] However, Pannenberg chooses to focus his critique of relativity primarily on the issue of measurement.[51]

47 Even Eddington is quite clear on this point, that the so-called "triumph of exact science" is simply in "establishing a numerical connection" between various systems of measurement. "The essential point is that, although we seem to have very definite conceptions of objects in the external world, those conceptions do not enter into exact science and are not in any way confirmed by it. Before exact science can begin to handle the problem they must be replaced by quantities representing the results of physical measurement"; Eddington, *The Nature of the Physical World,* 245. The object *per se* is extraneous to the exact sciences since they have no point of access to that object other than via the measurement of those characteristics (mass, volume, density etc.) which are relevant to the current equation. The object cannot enter into scientific equations until it has been replaced by a register of measurements.

48 Polkinghorne, *Belief in God in an Age of Science,* 52.

49 Ibid. 53.

50 "Two readings may be *equal,* but it is meaningless to inquire if they are *identical;* if then the [object] is a bundle of pointer readings, how can we ask whether it is continually the *identical* bundle?"; Eddington, *The Nature of the Physical World,* 248, cf. also 240–63.

51 Pannenberg identifies this first line of argument (the insufficiencies of relativity theory) but instead of following through on this insight he concentrates on the second point (measurement). We will return below to this first line of argument and see if we can strengthen this aspect of Pannenberg's critique.

3.3.1 Part Presupposes the Whole

Pannenberg's first attempts to engage seriously with relativity theory came very late in his career. As late as 1998, Polkinghorne could wonder at Pannenberg's failure to engage with relativity, despite (or perhaps due to) a range of tangential remarks in his *Systematic Theology*[52]. Polkinghorne correctly noted that it "is certainly not possible to talk seriously today about space without taking relativity into account", and yet Pannenberg's work in this area has displayed instead "a preoccupation with a rather old-fashioned concept of inertia".[53] Yet in his recent work, such as his 2005 essay "Eternity, Time, and Space", Pannenberg turns his attention to relativity theory. He begins his critique by appealing to Kant's analysis of space and time as found in Kant's *Critique of Pure Reason.* According to Pannenberg, Kant argued that "all of our conceptions of spaces and of specific times presuppose a prior intuition of *space as an infinite whole* and of *time as an infinite whole,* because different spaces can be distinguished and related only within some prior space, and different times are conceived as parts of one and the same time".[54] Thus divisions in space (centimetre, kilometre, lightyear etc.) can only occur *as divisions* within a preconceived *unity* of all space. Likewise divisions in time (minute, year, millennium etc.) can only occur *as divisions* within a preconceived *unity* of all time. "Therefore, some infinite and undivided whole of space and time is a prior condition in forming any conception of spatial or temporal units".[55] Importantly, these concepts of space and time as infinite wholes therefore take precedence over, and are "prior to all geometry, because the spatial and temporal units needed for measurement are themselves parts of space and of time, which precede as infinite wholes all notions of partial spaces and times".[56]

Although Pannenberg recognizes that Kant's argument was developed as part of his analysis of human consciousness, the validity of the thesis cannot be limited solely to the field of the human consciousness

52 Cf. e.g. Pannenberg, *Systematic Theology,* ii. 81, 88, 89, 91, 93, 111, 154–5.

53 Polkinghorne, *Belief in God in an Age of Science,* 81–2.

54 Pannenberg, "Eternity, Time, and Space", 98, emphasis added. Cf. also Pannenberg, *Systematic Theology,* i. 406–7.

55 Pannenberg, "Eternity, Time, and Space", 98.

56 Ibid.

of space and time.[57] Indeed, Pannenberg points to Kant's dependence on the work of Samuel Clarke, who applied this view to the "objective content" of experience. "Clarke argued that in all spatial division and composition some infinite and undivided space is presupposed, a space within which operations of division and composition become possible", and Clarke considered this to be "the space of God's omnipresence".[58] At this point, Pannenberg notes that, "Even the relativistic concept of space-time works with measurement, which needs units of measurement that are conceivable only within some prior, infinite, and undivided space. ... [Therefore, a]lthough relativity has an impact on the philosophical issues of space and time ... still it does not completely reconstitute our notions of space and time".[59] For Pannenberg, such a concept of the measurement of discrete parts requires a preceding concept of the absolute whole.[60]

Two points need to be made here. First, as we have seen Whitehead's process philosophy (and particularly the concept of time he puts forward) and the relativistic "theory of the manifold" which adopt the four-dimensional manifold of Minkowski space-time take up opposite

57 Ibid.

58 Ibid.

59 Ibid. 99. This criticism "applies to the definition of time in modern physics, which follows the Aristotelian tradition inasmuch as it starts with the measurement of time, so that a unit of movement that serves as a standard—light in relativity theory—is basic"; Pannenberg, *Systematic Theology*, ii. 93.

60 While the precedence of the whole over discrete parts is a reoccurring theme in Pannenberg's work, he misapplies it here to relativity. Cf. e.g. Wolfhart Pannenberg, "The Problem of the Absolute", in id., *Metaphysics and the Idea of God* (Grand Rapids, MI: Eerdmans, 1990), 22–42 at 22: "There is no metaphysics without the idea of the unity of reality. And because the unity of the real, understood as world or *cosmos*, is a unity that consists in the order of the many (i.e., of many individual things), the unity of the world raises the question of a ground, a foundation, that is able to order the many and hold it together as a unity". The problem is that the Minkowski universe does indeed presuppose such an underlying whole in the form of the absolute and invariant space-time manifold. Cf. also Wolfhart Pannenberg, "Theology and the Categories 'Part' and 'Whole'", in id., *Metaphysics and the Idea of God* (Grand Rapids, MI: Eerdmans, 1990), 130–52.

positions in the philosophical debate on time. And so it is disconcerting to see Pannenberg attempt to dismiss them both using the same argument based on measurement. When Pannenberg finally comes to engage with relativity, he simply returns to his old idea of an underlying absolute or whole which ensures the unity of reality—a point of critique he used against process philosophy, and the theology which has arisen from it. The theoretical tension between "part" and "whole" is a concern that can be traced through Pannenberg's larger corpus. Yet the appeal to the whole as that which underlies particular instances, and the whole's status as a "superior reality" was developed in Pannenberg's early critique of what he saw as the "atomism" of process philosophy.[61] It was in his encounter with process philosophy that Pannenberg took issue with what he understood to be Whitehead's claim of the primacy of "parts" over the "whole": that "discrete emergent 'occasions' or elementary events are the ultimate realities".[62] Such a bottom-up approach implies for Pannenberg "an atomistic ontology",[63] and thus he finds Whitehead to have fallen victim to "the logical aporias of every atomistic metaphysics":[64] that "*an encompassing unity must be presupposed* if atoms are to be conceivable as unities at all".[65] Thus, rather than progressing from the initial "whole", which then guarantees the reality of its component "parts", "Whitehead proceeds from the ontological priority of discrete events or their components",[66] so that "it is always the individual event itself which must integrate into a whole".[67] "No atomistic idea can explain the unity of the many—either in the whole which is the world, or as displayed in the more or less enduring entities and forms".[68] Thus in the space-time

61 Cf. esp. Pannenberg, "Atomism, Duration, Form", and Wolfhart Pannenberg, "Being and Time", in id., *Metaphysics and the Idea of God* (Grand Rapids, MI: Eerdmans, 1990), 69–90.

62 Pannenberg, "Atomism, Duration, Form", 114.

63 Ibid. Pannenberg takes issue with Whitehead's claims that "the ultimate metaphysical truth is atomism", and that his philosophy represents "an atomic theory of actuality"; Whitehead, *Process and Reality*, 35, 27.

64 Pannenberg, "Atomism, Duration, Form", 116–17.

65 Ibid. 117, emphasis added.

66 Ibid. 116.

67 Ibid. 119.

68 Pannenberg, "Being and Time", 72.

model, Pannenberg diagnoses the same error at work in Whitehead's process thought.

And yet, second, it should be clear at this stage that the four-dimensional Minkowski universe does indeed place a radical primary stress upon the absolute wholeness of space-time and the subsequent relativity of all partial units of measurement. Minkowski's achievement was to reunify the disparate aspects of relativistic space and time into a singular, four-dimensional, invariant object. Measurement of that object will differ in and of itself, since the units of measurement have become arbitrary, but the absolute whole is still ontologically primary—it is measurement, and the units of measurement, which have become relative, dependent both upon one's particular frame of reference and the contractions and expansions of space-time between those differing frames.[69] Pannenberg critiques relativity on this point by supposing that it still operates with the preconceptions of classical physics, that relativity theory (or the geometry of Minkowski space-time and relativistic mechanics) in principle *first* assume units of measurement and then only subsequently postulate the absolute as the combination and compilation of those units. On the contrary, we have seen that in Minkowski space-time it is the absolute that takes ontological precedence: measurement is an attempt to interact with and categorize that absolute in an "exact" (albeit irretrievably relativistic) manner. In broad terms, the difficulty which has polarized the proponents of process philosophy and special relativity is that process theory assumes a bottom-up approach whereas the Minkowski universe takes a top-down approach. A misunderstanding of this distinction leads to a critique on Pannenberg's part which fails to sufficiently impact on the philosophy of special relativity.

However, in his discussion of the wholeness of time, Pannenberg also introduces an additional distinction, stressing that the absolute space-time of the four-dimensional manifold is not the infinite whole of time by which theology seeks to express divine eternity[70], since the "absoluteness of space-time ... replaces the absoluteness of the eternity that con-

69 See the discussion in Eddington on the impact of the FitzGerald contraction and the conflict between relative and absolute quantities; Eddington, *The Nature of the Physical World*, 16–30, 34–8.

70 Pannenberg, "Eternity, Time, and Space", 99.

stitutes time".[71] From a theological perspective it is easy to understand why Pannenberg want to maintain such a distinction. If we assume that the absolute four-dimensional manifold is coterminous with "the infinite space of God's omnipresence", then the being of God and the being of creation co-inhabit the same space-time and pose a series of pantheistic problems. Pannenberg clearly seeks to avoid this consequence, and therefore argues that "space-time is not eternity": "A pantheistic view of space-time suggests itself only if the undivided nature of infinite space (and time) is not distinguished as it should be from all geometrical descriptions of space and time".[72] But how are we to interpret this move? There are two directions here in which Pannenberg may be moving:

3.3.2 Ontological Critique

On the one hand, Pannenberg's argument appears to be reintroducing an ontological split, or metaphysical distinction between the eternal, divine realm, and the absolute space-time of relativity theory. This impression is strengthened when we read Pannenberg's assertions that, "Space-Time as a geometrical concept describes the comprehensive field of all *finite phenomena,* especially of matter and masses", and that "the space-time concept" is to be connected "with the occurrence of *material* phenomena, of masses, in the universe".[73] The suggestion here is that the world of material phenomena (or in the terms of Minkowski space-time: the four-dimensional manifold) is not to be confused with God's eternity, with the eternal space of God's omnipresence. However, the difficulty here is that omnipresence implies *spatial relations,* thus a certain and necessary overlap between God's "space" and the "space" of space-time (if one assumes that divine omnipresence signifies the divine presence in and through all *created space*). Pannenberg is aware of the problem and agrees that "Ideas of God's omnipresence are inexplicable without some connection with the concept of space, but it has to be a concept of space and time that is *different* from geometrical space and time, prior to

71 Pannenberg, *Systematic Theology,* ii. 93 n. 241.
72 Pannenberg, "Eternity, Time, and Space", 101.
73 Ibid. 100–1, emphasis added.

all measurement, if the distinction between God and the world is to be observed".[74]

The problem for Pannenberg here is how to keep two theistic tenets distinct: that God is omnipresent, and that God is not coterminous with his creation (pantheism). To achieve this distinction, Pannenberg appeals to *different* forms of space and time, distinguishing between "the world of finite experience, the world of nature", and "God's eternity and omnipresence".[75] While this move is hardly surprising, it is worth noting that by doing so, Pannenberg appears to be reintroducing an ontological distinction between geometrical space and divine space, between the finite world and God's eternity. He re-establishes a metaphysical, ontological duality in order to defend a theological tenet: precisely the move he heavily criticized in Barth, Bultmann and Kähler, and the move he sought to avoid.[76]

By way of comparison, Jürgen Moltmann elegantly side-steps this problem by simply rejecting the difficult (and largely extra-biblical) concept of omnipresence[77], suggesting instead that in creation God's withdrawal creates a "God-forsaken place".[78] The advantage with such a move is that it secures the gift of otherness for God's creatures, distancing them from the being of God. If omnipresence could ever be predicated to God, then (according to the doctrine of *creatio ex nihilo*) this would be when

74 Ibid. 101, emphasis added.

75 Ibid.

76 Cf. Pannenberg, "Heilsgeschehen und Geschichte I", 218.

77 Though a detailed biblical analysis will not be presented here, cf. the extensive biblical witness to YHWH as "a God who hides himself" (Isa. 45:15), "turns away", "hides his face" and "abandons" his people (see e.g. Deut. 31:17–18; 32:20; 1 Chr. 28:9; 2 Chr. 12:5; 15:2; Neh. 9:28; Job 13:24; 34:29; Pss. 10:1; 13:1; 22; 27:9; 44:24; 55:1; 69:17; 78:56–64; 88:14; 89:46; 102:2; 104:29; 143:7; Isa. 45:15; 54:7; Ezek. 39:29; Mic. 3:4; as well as Tob. 13:6 and 2 Esd. 5:9 among many others).

78 Moltmann, *God in Creation*, 87. Though the concept of God's free self-limitation as a prerequisite for the gifting of otherness is not limited to Moltmann. In the science and theology discourse see also I. G. Barbour, *Religion in an Age of Science* (San Francisco: Harper and Row, 1990), ch. 6; and the detailed exposition by W. H. Vanstone, *Love's Endeavour, Love's Expense* (London: Darton, Longman and Todd, 1977).

creation did not exist.[79] In the very act of creation, in bringing into being that which is not God, God creates a "God-forsaken place", distinguishing Godself from that which is not God, and thus sacrificing any claim to *substantial* omnipresence.[80] While one may suspect here that Pannenberg is mistaking God's immanence for God's omnipresence, it is clear that Pannenberg is not eager to negotiate on this point of "faith", arguing simply that "we cannot follow Moltmann's thesis". "Opposed to this thesis is the truth that God is omnipresent in the space of creatures".[81]

3.3.3 Epistemological Critique

A second possibility is that Pannenberg intends here an *epistemological* distinction between God's eternity and the finite world. The stress would then fall upon Pannenberg's statement claiming "Space-Time as a geometrical concept",[82] that is, as a conceptual aid for understanding the world of finite phenomena, but one which is unable adequately to

---

79 In this context, "when" does not represent a temporal qualifier since created time is by definition part of creation. Here Pannenberg helpfully notes that there "is no *measurable* time and space without creatures. ... God created time and space as dimensions of the existence of the world of finite creatures"; Pannenberg, "Eternity, Time, and Space", 105, emphasis added.

80 The issue here is one of *substantial* omnipresence, i.e. where the *being* of God is coterminous with or overlaps the being of creation. Such a view was already critiqued by Thomas Aquinas, who noted (as part of his nuanced view on omnipresence) that by analogy a person need not fill the entire space of a room to be present in it. Rather "in virtue of its presence, a thing exists in everything within its field of view, so that everything in a house is said to be present even to a person not existing substantially in every part of that house. ... We talk of something being present to someone whenever it lies within his field of vision, even if in substance it is far away from him"; Thomas Aquinas, *Summa Theologicæ*, 1a, Q8, a3.

81 Pannenberg, *Systematic Theology*, ii. 89 n. 229. The discussion in the *Systematic Theology* makes it clear that for Pannenberg, the omnipresence of God is closely linked to God's omnipotence and omniscience in a substantial way, a move that can be traced back at least to Duns Scotus; cf. Pannenberg, "Archbishop Thomas Bradwardine", 174.

82 Pannenberg, "Eternity, Time, and Space", 100.

understand the space of God's eternity and omnipresence. This seems to be the case when Pannenberg speaks of "a different view of eternity" and when he adopts "the Plotinian idea of eternity as simultaneous presence and possession of the wholeness of life".[83] Pannenberg's concept of time (which we will examine in the following chapter) introduces a concept of eternity as "undivided infinite space and time",[84] an "omnitemporality"[85] which encompasses future, present and past[86] in contradistinction to the way in which "measurable" time is understood. However, if the issue is simply an epistemological difference in the conceptions used to understand time (that is, without a differentiated ontological basis), then such an approach would have difficulty in overcoming the pantheistic problem outlined above. Furthermore, it raises the question why such an epistemological distinction in the understanding of space and time has arisen between the exact sciences and theology. Just as Bergson and the German historicists argued that the natural sciences (with their strict focus on rationality) were unable properly to comprehend freedom or historical empathy, so too if the natural sciences are *incapable* of understanding space and time in the way theology can, does this not just establish (yet again) a safe harbour in which theology can operate, and a dual language methodology inaccessible to the epistemology or claims of the natural sciences?

The key here would be to display not the inability but the *unwillingness* of the natural sciences to operate beyond their own reductionistic scheme, and to plot out the precise boundaries where Minkowski space-time encounters its limitations. The goal of such a move would be to reject the all-or-nothing approach to epistemology, to critique "those who hold that exact science is all-sufficient for the description of the universe and that there is nothing in our experience which cannot be brought within its scope".[87] In this way, by displaying the limitations of relativistic mechanics, one could gain space for a more nuanced conception of time, which is, after all, Pannenberg's aim.

83 Ibid. 102.
84 Ibid. 105.
85 Ibid. 102.
86 Pannenberg, *Systematic Theology*, ii. 92–3.
87 Eddington, *The Nature of the Physical World*, 246.

3.4 Extended Critique of Space-Time

Pannenberg's decision to attack relativity on the basis of measurement provided an important insight into the positivistic foundation underlying the exact sciences, but remained largely tangential to the main structure and, above all, principle assertions made by the model of Minkowski space-time. If the intention is to reclaim space from relativity theory for other conceptions of time, then it may be more helpful to attack the claim that the theory of time presented in the Minkowski universe is in fact sufficient and completely adequate for understanding the temporal dimension of all that exists—a claim which seeks to render all other temporal systems obsolete. While Pannenberg was open to a critique of relativity on these terms[88], it largely remains an assertion in his work. What is needed instead is a significant analysis and critique on these terms, outlining the points where the Minkowski universe fails to account for insights into reality provided by fields outside the natural sciences. It is precisely this task that this section aims at, to pick up Pannenberg's assertion and attempt to extend his critique of relativity along these lines. In short: If the intention is to reclaim space from relativity theory in order to make room for other, competing and more nuanced theories of time, can it then be shown that there are aspects of time and reality which relativistic mechanics fail adequately to reflect?

We might assume that Pannenberg's appeal to an omnitemporal eternity and a state of divine transcendence are aimed precisely at showing the limitations of relativistic mechanics. Such a move is hardly surprising for a theologian. Yet it fails to be a valid point of argument since it is the existence of a transcendent dimension which relativity theory calls into doubt. Nor should it be assumed that schools of theology and philosophy have always been critical of space-time ontologies. There is a significant strand which has accepted and supported the assumptions and conclusions of Minkowski's work and the later physics which was built upon it, and have sought to incorporate them into philosophy and theology.[89]

88 Pannenberg, "Eternity, Time, and Space", 97–8.

89 There is no space here to provide a detailed overview of space-time theory in philosophy and theology. Rather I have chosen one example from each field: Donald Williams in philosophy, and Alan Torrance in theology.

In his 1951 essay *The Myth of Passage,* the Harvard philosopher Donald C. Williams sought to support the manifold theory of time within the philosophical discourse.[90] He refers to the standard concepts of space-time theory we have already seen. He critiques popular views which see objects "moving" through time, which relegate these objects to a "limbo" of non-being as soon as they pass from the present into the past. He notes that "reality" has come to refer solely to the present moment, since future events are not yet real and past events have ceased being real.[91] Such views, he argues, have led "to the assumption that the totality of existence is to be identified with occupancy of an infinitely non-extended, indivisible 'now'".[92] In contrast, Williams' goal is to assert an understanding of reality as extended in four-dimensions. In line with Minkowski's work, Williams states that four-dimensional entities neither move through time nor pop in and out of existence along with the moving plane of the present. "Nothing can 'move' in time alone any more than in space alone, and time itself cannot 'move' any more than space itself. ... [Each] of us proceeds through time only as a fence proceeds across a farm: that is, parts of our being, and the fence's, occupy successive instants and points, respectively. ... [Passage] is the mere

90 D. C. Williams, "The Myth of Passage", *The Journal of Philosophy,* 48/15 (1951): 457–72.

91 The widespread dependence on this view is at the basis of many of our problems explaining the "real" but "not yet" character of Christian eschatology and the kingdom of God; cf. Christoph Schwöbel, "The Church as a Cultural Space: Eschatology and Ecclesiology", in John Polkinghorne and Michael Welker (eds.), *The End of the World and the Ends of God: Science and Theology on Eschatology* (Harrisburg, PA: Trinity, 2000), 107–23, esp. 111–12.

92 Alan J. Torrance, "Creatio Ex Nihilo and the Spatio-Temporal Dimensions, with special reference to Jürgen Moltmann and D. C. Williams", in Colin E. Gunton (ed.), *The Doctrine of Creation: Essays in Dogmatics, History and Philosophy* (Edinburgh: T & T Clark, 1997), 83–103 at 96, attributing these ideas to Williams rather than Einstein and Minkowski upon whom Williams' work draws. This is likely due to Torrance's dependence upon J. J. C. Smart's summary of Williams' argument in his article "Time" in Paul Edwards (ed.), *Encyclopaedia of Philosophy* (London: Collier-Macmillan, 1967), viii. 126. See Torrance, "Creatio Ex Nihilo and the Spatio-Temporal Dimensions", 94 n. 32.

happening of things, their strung-along-ness in the manifold."[93] Once we learn to perceive entities as invariant four-dimensional solids, to "accept realistically the four-dimensional fabric of juxtaposed actualities", and "believe that the universe consists, *without residue,* of the spread of events in space-time" then, Williams believes, "we can dispense with all those dim non-factual categories which have so bedevilled our race: the potential, the subsistential, and the influential, the noumenal, the numinous, and the non-natural"[94], as well as that particularly long-lasting and troublesome concept, "becoming".[95] Overly-eager theologians have picked up Williams' work (or, more precisely, its underlying foundation in special relativity and Minkowski space-time theory) and imported it wholesale into their theological systems; yet can it in fact be demonstrated that Minkowski space-time is sufficient to explain the actuality of the universe 'without residue'? Can this be done without appealing to a transcendent plane, but merely by referencing so-called 'mundane phenomena'? If this can be achieved, then not only can new space be created for alternate theories of time, but a guide will be produced which outlines the criteria which such an alternate theory will have to meet if it is to be considered sufficient.

3.4.1 Space-Time as the Totality of All that Is?

Alan J. Torrance picks up the insights from Williams' "profoundly important"[96] essay, and declares: "A theological doctrine of *creatio ex nihilo* should ... interpret God as creating space-time *per se,* the entire spatio-temporal nexus, *the totality of all that is,* together with the ordering and inter-relating matrices constitutive of the totality of created identities. In other words, God creates the space-time manifold *ex nihilo*".[97]

The great strength of relativistic mechanics is its elegant ability to deal with the motion of individual particles. However, this is also the

93 Williams, "The Myth of Passage", 463.
94 Ibid. 458, emphasis added.
95 Ibid. 463.
96 Torrance, "Creatio Ex Nihilo and the Spatio-Temporal Dimensions", 93 n. 30.
97 Torrance, "Creatio Ex Nihilo and the Spatio-Temporal Dimensions", 98–9, non-Latin emphasis added.

point where it finds its limitation. As a system of mechanics, it is highly structured to investigate *substantial entities,* such as atoms and particles. A major problem arises, however, if we seek either to apply relativistic mechanics to non-substantial realities or submit ourselves to the tenets of scientism and declare that nothing in addition to substantial entities exist, that nothing exists apart from that which can be measured with relativistic mechanics.[98] While Torrance's declaration seems to align him with the latter alternative, Williams himself was not prepared to make such a broad statement erasing the non-substantial from the universe. In guarded terms Williams writes, "I am arguing here, not that there is nothing outside the natural world of events, but that the theory of the manifold is anyhow literally true and adequate to that world".[99]

The list of non-substantial realities (time, love, data, etc.) and the scientific disciplines devoted to non-substantial realities (behavioural sciences, information sciences, etc.) are numerous. But to unpack this issue, let us look closely at two examples of non-substantial realities which cannot be integrated easily (if at all) into the geometry of Minkowski space-time: information and pain.[100]

---

98 This danger certainly predates Torrance and Williams and the modern debates on scientism. In his Gifford lectures, Eddington raised the overwhelming tendency of the sciences (and human perception in general) to be preoccupied with substantial reality. "So strongly has substance held the place of leading actor on the stage of experience that in common usage *concrete* and *real* are almost synonymous. Ask any man who is not a philosopher or a mystic to name something typically real; he is almost sure to choose a concrete thing. Put the question to him whether Time is real; he will probably decide with some hesitation that it must be classed as real, but he has an inner feeling that the question is in some way inappropriate and that he is being cross-examined unfairly. ... In leading us away from the concrete, science [should be] reminding us that our contact with the real is more varied than was apparent to the ape-mind, to whom the bough which supported him typifies the beginning and end of reality"; Eddington, *The Nature of the Physical World,* 264–5.

99 Williams, "The Myth of Passage", 458.

100 I am purposefully avoiding a theological discussion about the nature of God here, whether the being of God should be classified as a substantial reality within Minkowski space-time or as a non-substantial reality external to it (or some [pan{en}theistic?] mixture of the two). Instead, we can ask whether simple, less contested aspects of what we would consider mundane reality fit neatly

Physics works well when dealing with point events, but things become more complex with macro-level objects, especially when helpful reductionisms are set aside. Yet apart from the complexities of dealing with macro-level objects, Whitehead also suggested that physics was unable to deal adequately with information or "pattern" that arises on the scale of macro-level events and their interactions.[101] Whitehead begins with the following thesis: "An event is the grasping into unity of a pattern of aspects".[102] Whereas Minkowski space-time postulates a static, block universe which *exists* rather than properly *endures* (since endurance implies a dynamic temporality), Whitehead suggests such a dynamic notion of endurance where events are effective in producing successive events and effects upon their surrounding environment: "The *effectiveness* of an event *beyond itself* arises from the aspects of itself which go to form the prehended unities of other events. ... The event constitutes a patterned value with a permanence inherent throughout its own parts; and by reason of this inherent endurance the event is important for the modification of its environment".[103]

For Whitehead, it is here "in this endurance of pattern that time differentiates itself from space". The pattern is spatially extended, and the endurance of that pattern introduces temporality. Whitehead takes pains to stress the developmental change that is bound up in the concept of endurance: by endurance he does not mean the static, four-dimensional invariant extended through Minkowski space-time (that is, with a unity of *material identity*), but "a *pattern* which is exhibited in the prehension of one event [which] is also exhibited in the prehension of those of its parts".[104]

---

into the relativistic mechanics of the Minkowski universe.

101 Whitehead's intention is to remedy this deficit by postulating his own "organic" theory of nature. While Whitehead understands his organic theory as "a basis for a thoroughgoing objectivism" (Whitehead, *Science and the Modern World,* 119)—thus placing it within the class of objective theories of time—his tendency to use active verbs ("grasping") and qualitative expressions ("unity", "pattern") seems to link him here still with aspects of the subjective. And yet the subject, and extent of that subjectivity is still unclear in his early work.

102 Whitehead, *Science and the Modern World,* 119.

103 Ibid. emphasis added.

104 Ibid.

Whitehead goes on to ask: If "each enduring object discovers in nature and requires from nature a principle discriminating space from time",[105] then do "all enduring objects discover the same principle of differentiation of space from time ... [or may they not] vary in [their] spatio-temporal discrimination"?[106] Whitehead notes that pre-relativity, time would have had exactly the same meaning when comparing the endurance of two moving objects, but that now the "meanings of space and time ... [are no longer] identical from one object to another".[107] Thus the loss of a "definite meaning to the geometrical relations" of space also required physics to restate such fundamental principles as gravity.[108] As for temporality, this lack of "reference to a definite space-time system" has even robbed us of our experience of simultaneity, and this, for Whitehead, constitutes one of his most pressing concerns regarding the theory of relativity, which he "cannot reconcile ... with the given facts of our experience as to simultaneity and spatial arrangement".[109]

---

105 Ibid.

106 Ibid. 120.

107 Ibid.

108 Ibid. 121–2.

109 Ibid. 122. While Whitehead correctly notes the inability of physics to deal with pattern and thus keep space and time separate (as consciousness demands), the issue in relativity is one of *measurement,* not of "multiple space-time systems" (as correctly noted by Pannenberg and Craig above). Minkowski postulated space-time as common to all observers and thus absolute. The importance of Minkowski's unification of space and time into a unified four-dimensional manifold was that it avoided such an inescapable multiplicity of space-time systems: the relativity of various frames of reference, in relativity theory, is a relativity only in *measuring* the same absolute, invariant four-dimensional manifold. As Whitehead himself admits: "Einstein would probably reject the theory of multiple space-time systems which I have been expounding to you. He would interpret his formula in terms of contortions in space-time which alter the invariance theory for measure properties"; ibid. Here, Whitehead displays his interest not in responding to interpretations of relativity theory, but in producing his own organic interpretation which would "reclaim" and maintain a distinction between the temporality and extensiveness of events (ibid. 118) which Minkowski space-time does not allow. Even Eddington expresses his concerns regarding "the rather heterodox views of Dr. Whitehead on relativity" and the perception that Whitehead's work at that time had failed adequately to represent or engage

Whereas Minkowski space-time accounts for the permanence and spatiotemporal extension of individual particles, Whitehead is correct to note this difficulty in accounting for the organic interaction of particles in larger systems. The atoms themselves may express a degree of permanence, yet over the life-span of an organism they will constantly be exchanged even though the overall pattern of the organism (the event's essential individualization) is retained. Thus Whitehead presents us with a mutually influencing process of *endurance* and *reiteration.* He stresses that in his "organic theory, a pattern need not endure in *undifferentiated sameness* through time. [Indeed the] pattern may be essentially one of *aesthetic contrasts* requiring a lapse of time for its unfolding. ... Thus the *endurance of the pattern* now means the *reiteration of its succession of contrasts*".[110]

While Whitehead's Platonic tendencies are visible here in his claim that the "concept of internal relations requires the concept of substance as the activity synthesizing the relationships into its emergent character",[111] we could also make a similar argument about the distinction between matter and information (and the inability of relativistic mechanics to account for the distinction) by appealing to a less material phenomenon: language.

Ever since J. L. Austin's groundbreaking treatise, *How To Do Things With Words,* there has been growing appreciation for the power of the performative utterance. The ability to use language not only in the descriptive or indicative sense (merely for stating facts), but also in a performative and forceful way[112], and in the transmission of information, highlights an entire range of action and force disconnected from the way these concepts are described in physics. At very best, one could try to assimilate these speech acts into relativistic mechanics via a concentration on the simple mechanics of sound: the uttering of noises and the transferral of energy as waves through air. Yet this limits language to the mere

---

with relativity theory as Einstein and Minkowski had developed it; Eddington, *The Nature of The Physical World,* 146–7, though he does attempt to make more conciliatory moves at 242.

110 Whitehead, *Science and the Modern World,* 133, emphasis added.

111 Ibid. 123.

112 J. L. Austin, *How to Do Things With Words* (2nd edn.; Cambridge, MA: Harvard University Press, 1975).

phonetic act of sound production, and scrubs away all meaning carried by those sounds, failing to engage the words at even the phatic or rhetic levels[113] let alone appreciating their elevated locutionary, illocutionary or perlocutionary force[114]. Even if we were to simplify and reduce language to a basic indicative level, relativistic mechanics would completely lose the information that is contained within those waves of energy. There would be no way to register or represent the information or significance tied to those sounds.

In all fairness, we could admit that Whitehead's concentration on pattern and the endurance of complex systems, as well as the practical information aspects of language should not really be taken as a strict critique of relativistic mechanics, since mechanics by its nature is not designed to deal with information transfers. Yet this is precisely the point. Neither should we accept those who, with a narrow scientistic perspective, would simply declare complex systems and information to be non-existent simply because relativistic mechanics is incapable of representing them.

The inability of information and pattern to be incorporated into models of relativistic space-time already calls into question the claim that such a model is adequate for understanding "the totality of all that is" (to adopt Torrance's expression). But we can go further and include here also basic and universal human affects. Pain presents a good practical example of such universal affects, and clearly displays, again, the limits of relativistic mechanics.[115] Even if we were temporarily to set aside psychological pain (i.e. pain which arises from a relation to consciousness, such as sorrow and anguish), and concentrate solely on physiological (cutaneous, somatic, visceral, and neuropathic) pain, then we would be faced with a dimension of reality testified to by vast swathes of biological life but which can hardly be accommodated within the model of the four-dimensional universe.

113 Ibid. 92–3.

114 Ibid. esp. lectures 8, 9 and 10. See also Mikhail Bakhtin, *Speech Genres and Other Late Essays*, trans. V. W. McGee (Austin: University of Texas Press, 1986), 60–102.

115 Theologians and romantics might prefer to discuss love rather than pain. Yet love is neither universally expressed or experienced, nor can it be quantified neuroanatomically.

At the human level, the issue of pain, and particularly suffering, becomes acute. It might be argued that while physiological pain is poorly addressed in the Minkowski universe, psychological pain actually presents some parallels to the general philosophy (if not the mathematics) of invariant four-dimensional objects. When we speak metaphorically, true human sorrow and suffering are not discrete emotions which pop into the present, possess a short span of existence, and then disappear with the "passing" of time. In this sense, the "myth of passage" applies as much to human anguish as it does to the temporal "flow" of concrete objects. A flowing conception of time quickly forgets suffering and relegates it to past non-existence, yet anguish enters into human life, stains it, numbs it, and refuses to be removed. It becomes part of one's extended four-dimensional existence and requires a regenerative, healing act of God to extract it, if at all, or transform it.[116] In this regard, the language of the Minkowski manifold is deeply appropriate to the reality of human suffering since it provides a framework for the permanence and radical embeddedness of suffering in this sense. Yet it is *only* in this metaphorical sense. As soon as we try to apply relativistic mechanics in all its mathematical rigour to the reality of suffering, we hardly know how to begin. The metaphor falls apart and we are faced again with the inadequacy of relativistic mechanics to account for a very real dimension of reality. While the extension of space-time offers a better *metaphor* for the embeddedness of human suffering (over against the transitoriness of a "temporal flow"), this can in no way be elevated above the level of a simple metaphor, precisely due to the limitations of Minkowski's model. The model of space-time mechanics is simply not suited (and not meant) to deal with these aspects of reality.

Thus it is worth repeating that a critique along these lines is not levelled strictly at Minkowski's work *per se,* nor even at relativistic mechanics, but at the *misappropriation* of this work in philosophy, theology, and even physics itself, in the attempt to portray relativistic mechanics as adequate for describing the entire universe "without residue", as reflecting "the totality of all that is". The behavioural and information sciences, and the examples we have looked at here are just a small selection that point

116 Cf. the treatment in C. S. Lewis, *The Problem of Pain* (London: Bles Centenary, 1941).

to this inadequacy. Yet the neurobiologist or neuropsychologist might object that physiological pain (let alone psychological suffering), however real, is a poor test of measurable reality since (in human beings at least) even physiological pain is a product of consciousness and not a component of the physical world.[117] However, this brings us to a second major deficit of the space-time model.

3.4.2 The Abandonment of Subjectivity

When we sketched out the development of the geometry of Minkowski space-time above, we noted that in the previous Cartesian system of analytic geometry, the origin of the spatial diagram was marked by an origin, a meeting of axes that we could call "myself". This point was used as the basis for all subsequent measurements: the things that *I* saw occurring "now" *I* could place along my own "now line", and events to *my* left or right could be plotted in accordingly. In short, this system was still closely connected with the individual's own perception of events, and in this respect retained an aspect of subjectivity.

But in the shift from this rather Cartesian system to Minkowski's four-dimensional manifold, we first extended the "myself" point into a four-dimensional line, together with the "now" line which represented what we believed to be happening "now". Yet we noted then that the space-time model finally erases both the "now" line as well as the "myself" line since neither was inherent or "essential to the world".[118] Having made this step, Minkowski gains an objective conception of the absolute structure of space-time. Yet in the same move consciousness and the knowing subject are also removed from the theoretical universe, together with all that consciousness entails.

117 The neural sciences speak only of nociception and the stimulation of nociceptors (sensory neurons specialized for the detection of noxious stimuli). "Pain" *per se* is seen as a product of consciousness, created by the brain's interpretation of stimuli received via the body's nociceptors. Cf. here E. R. Kandel, J. H. Schwartz and T. M. Jessell (eds.), *Principles of Neural Science* (4th edn.; New York: McGraw-Hill, 2000), 472–91.

118 Eddington, *The Nature of the Physical World,* 56.

Rather than representing a problem for physics, this was in fact a significant achievement. In its search for absolute objectivity and a rigorously self-contained domain of study, physics has constantly and necessarily sought to remove any reference to the "knowing subject" that stands outside the self-referential cycle of measurement.[119] Physics is not at all anxious to pursue the question of consciousness, or of what consciousness knows, and from its own point of view this is entirely justified: that physical matter in some indirect way comes within the scope of consciousness is not a fact that bears any use for a theoretical model in physics: "We cannot embody [consciousness] in a differential equation".[120] However, this move leads to significant difficulties when particular aspects of reality are open to consciousness but not to theoretical physics, or when one seeks to account for the very existence of consciousness itself.

The problem only arises when we take the four-dimensional manifold of the block Minkowski universe as completely sufficient for understanding the *complexity* of nature and time, even though it fails to account for

---

119 Eddington provides a detailed example of the self-referential and cyclical method used and desired by the exact sciences, taking as his example the necessary steps in providing a complete explanation of gravitation: "Einstein's law in its analytical form is a statement that in empty space certain quantities called *potentials* obey certain ... differential equations. ... The next question is, What are potentials? They can be defined as quantities derived by quite simple mathematical calculations from certain fundamental quantities called *intervals*. ... What are intervals? They are relations between pairs of events which can be measured with a *scale* or a *clock* or with both. ... What are scales and clocks? A scale is a graduated strip of *matter* ... What is matter? ... Confining ourselves to mechanics ... matter may be defined as the embodiment of three related physical quantities, *mass* (or energy), *momentum*, and *stress*. What are 'mass', 'momentum', and 'stress'? It is one of the most far-reaching achievements of Einstein's theory that it has given an exact answer to this question. They are rather formidable-looking expressions containing the potentials and their first and second derivatives with respect to the co-ordinates. What are the potentials? Why, that is just what I have been explaining to you!"; Eddington, *The Nature of the Physical World*, 240–63 at 252–4. "By the ingenious device of the cycle physics secures for itself a self-contained domain for study with no loose ends projecting into the unknown"; ibid. 255.

120 Ibid.

essential aspects of the reality it seeks to represent.[121] It has already been noted that physical models of time are unable to deal adequately with the experiential concept of "the present"[122]; and since the universe of special relativity exists as an invariant block, temporal modes (as well as temporally influenced conditions such as "becoming") are rendered meaningless. This approach has entered into the philosophy of time, with the result that the so-called

> philosophers of the manifold hold that the flow of time or human advance through time is an illusion. They argue, for example, that words

---

121 Here we come upon a striking parallel to Spengler's harsh division between the historical "deed" and a-historical religion, understood as a shallow directedness towards the purely transcendent and thus away from "actuality". In both cases we are dealing with a form of reductionism (the former physical, the latter historical) which seeks to segregate its own field from its opposite, and in so doing to declare its own field coterminous with the bounds of reality/actuality. "The other world of historical actuality lying round [Jesus] was to him as something sham, alien, void of significance. That 'He' would now come and make an end of this unreal reality was his magnificent certainty. ... But when Jesus was taken before Pilate, *then the world of facts and the world of truths were face to face in immediate and implacable hostility.* ... In the famous question of the Roman Procurator: 'What is truth?' *lies the entire meaning of history,* the exclusive validity of the deed. ... The silence of Jesus answers this question by that other ... *What is actuality?* For Pilate actuality was all; for Him nothing"; Spengler, *The Decline of the West,* 287–8. From this perspective, religion (understood as a shallow directedness towards the purely transcendent, and thus away from historical "actuality") is seen as a rejection and negation of the historical process, and as such "true religion" for Spengler must point to an a-historical existence: "indeed, pure religiousness could never stand up against history and the powers of history, or sit in judgement on active life; or if it does, it ceases to be religion and is subjected itself to the spirit of history"; ibid. 288.

122 Cf. Paul Davies, *About Time: Einstein's Unfinished Revolution* (New York: Touchstone, 1996), 65. See also Polkinghorne, *Belief in God in an Age of Science,* 68: The "current equations of physics provide no lodging for the concept of the present moment. The time *t* appears as a physical parameter but there is nothing that identifies $t=0$ with 'now'. ... It's inability to reproduce so fundamental an aspect of human experience is to be interpreted as indicating the incompleteness of the scientific account rather than the illusory character of our experience".

> such as past, future and now, as well as the tenses of verbs, are *indexical expressions* that refer to the act of their own utterance. Hence the alleged change of an event from being future to being past is an illusion. ... Past and future are not real predicates of events ... and change in respect of them is not a genuine change.[123]

Torrance also follows this line, laying blame for our confusion regarding the nature of time and the universe on the subjectivity and system of tenses inherent in Indo-European languages: "'There will be a train strike next week!' [should] more properly be 'There is (first form [non-temporally-indexical] of the verb "to be") a train strike which is temporally located the week after this statement is (second form [indexical] of the verb "to be") being made!' A still more 'objective' non-indexical form would affirm the reality of the train strike and provide its spatio-temporal location simply by utilizing a series of space-time co-ordinates which made no appeal to the location of the affirmation of the existential statement itself".[124] The unwieldy awkwardness of this process, and the focus of its supporters on tense as a human construct, suggests that physics' difficulty with the present is in fact a second order problem grounded on its inability adequately to deal with human consciousness and its products.

In short, if relativistic mechanics cannot accommodate basic and fundamental aspects of human experience (such as pattern, information, language, or consciousness), then this should prompt us to appreciate the shortcomings and limitations of the theory and to search for complementary theories, rather than simply to dismiss these aspects as somehow illusory.

---

123 Smart, "Time", 126, emphasis added.

124 Torrance, "Creatio Ex Nihilo and the Spatio-Temporal Dimensions", 97. Note that modern German (and modern English to a lesser degree) is already making this transition by abandoning the future tense in favour of the present tense together with a temporal modifier, so that in German this sentence could read: *Es gibt nächste Woche Streikmaßnahmen bei der Bahn.*

3.5 Summary

This chapter examined the leading theory of time within the western natural sciences and, by presenting an examination of its development, sought to address the general confusion surrounding the assumptions, preconceptions and consequences of relativity theory, aspects of which were evidenced in the work of Wolfhart Pannenberg.

An analysis of Minkowski space-time was necessary here since the physical sciences present the *de facto* leading temporal theory. As a consequence, any academic discussion of time must first account for and engage with this dominant model. This is not to suggest that no other temporal theories are possible, but that space must first be made within the discourse if other competing theories are to be developed—especially if both theories attempt to speak to the same temporal reality (that is, they reject the two-language hypothesis which neatly segregates the truth claims of natural science from those of the human and divine sciences). Pannenberg has felt that the temporal aspects of the theory of relativity have failed to accommodate certain theological truths, and thus he has rightly called into question the dominance of relativity theory.[125] Yet rather than develop a narrowly theological concept of time which operates in exclusion from the truth-claims of the physical sciences and simultaneously excludes theology from the natural world, Pannenberg seeks a temporal theory that can account for material phenomena, especially those phenomena which are central to the Christian faith but excluded from the current framework of the natural sciences. Therefore, one of the aims of this chapter was to present Pannenberg's analysis of the insufficiency of relativity theory and to create space in that theory by displaying its limitations, simply in order that competing theories may have room to grow.

In accordance with the system of categorization developed in the previous chapter, Minkowski space-time can be classified as *linear* (since while the extension of space as finite but unbounded allows for a spatial recursion, it allows for "no bending by which Before ultimately becomes After"[126]), *objective* (since it exists external to human conscious-

125 Pannenberg, "Eternity, Time, and Space", 97.

126 Eddington, *The Nature of the Physical World*, 89.

ness, unaffected by consciousness or subjective perceptions of the flow of time), and *multireferential* (since interaction with and perception of the manifold is subsumed within an immense multiplicity of frames of reference). In the Minkowski geometry of space-time, the universe is the conglomerate of space, time and matter, an absolute, four-dimensional entity. Time is not a dynamic reality but is one incorporated dimension of an invariant object. The theoretical philosophy behind Minkowski's work leads to a series of consequences for a theoretical and theological understanding of time and history.

First, the space-time manifold is a *restricted* universe. Rather than presenting an open future of infinite possibility, the Minkowski universe is limited by the initial location of the particles which compose it. This restriction presents a degree of predictability for all particles which share proximity to each other and provides physics with a powerful predictive criterion for particles within a broad frame of reference. In theological terms, the restriction of the Minkowski universe, together with its unification of the dimensions of space and time, not only limits the openness of the future (providing the means for confronting and controlling the future's otherwise radical contingency) but also represents a challenge for the (admittedly) theistic concept of omnipresence.

Second, due to its strict limitation to four dimensions, the relativistic block universe presents an already complete, unified space-time entity which is invariant. Four-dimensional space-time is static, lacking a fifth or further (meta-)dimension in which to flow. And rather than postulating an "open future" into which entities grow, develop, change, and advance, the Minkowski universe is a conglomerate of already existent four-dimensional solids, stable objects which exist and extend as four-dimensional wholes without being formed in each new present moment only then to dissolve into the past. Expressed in theological terms by its supporters, creation would then be the coming into being of the entire four-dimensional block universe, and not merely the establishment of initial conditions which subsequently develop. The a-temporal block universe strips away all conceptions such as potentiality, development, and becoming.

Third, since those four-dimensional solids are invariant the Minkowski universe is a *determinate* universe. In respect to theology, Minkowski space-time would thus reject efficacious temporal intervention, be it human or divine.

Therefore, this model of space-time represents a starting point for a series of challenges and difficulties for theology, such as the call to repentance and the efficacy of supplicatory prayer. Furthermore, the determinate block universe presents problems for soteriology and *Heilsgeschichte,* particularly if that salvation history implies concurrent divine activity *in response* to creation. In addition, it rules out God's continuing providence and *creatio continua.*[127] Furthermore, it presents difficulties when dealing with the existence of evil and suffering in the universe, since if we agree that God creates "the entire spatio-temporal nexus, the totality of all that is, together with the ordering and inter-relating matrices constitutive of the totality of created identities"[128], then each person and each thing comes into being together with its multitudinous interrelations directly from the hand of God *ex nihilo,* and if the extension of one's four-dimensional existence leads a person into suffering, death and annihilation (regardless of whether this is under the forces of nature or at the hands of German Nationalism), then suffering, death and annihilation can only be that person's God-given, created, four-dimensional existence. Thus it becomes clear that from a theological perspective theodicy becomes the major challenge to all our temporal theories.

Despite long-held assumptions regarding the radical distinctions between the natural sciences and ancient theological traditions, the above examination of Minkowski space-time nevertheless revealed some first points of correlation and affinity with the biblical apocalyptic traditions. The radical integrity of Minkowski space-time (its unification of space, time, and matter into a single four-dimensional manifold) parallels the apocalyptic tradition's strong move to also join all time, history and reali-

127 Indeed Torrance critiques Moltmann's use of the expression *creatio continua* as well as Christoph Schwöbel's suggested distinction between *creatio continua* and *creatio continuata,* since "'continued creation' is no less guilty of suggesting that creation is extended in time than 'continuous creation'. Creation is simply not an infra-temporal event and nothing is to be served by conceiving of it as such"; Torrance, "Creatio Ex Nihilo and the Spatio-Temporal Dimensions", 98 n. 40.

128 Ibid. 98–9.

ty into a single totality, from its creation to its end.[129] They both postulate a block universe which completely encompasses that universe's component parts, and that block reality is invariant, with a determined course and end. Rather than encompassing change or development, change can only come through a process of complete destruction of that totality and its rebirth, either at the hand of God or in the heat and mass of the Big Crunch. Finally, the determined course of the reality totality allows for little efficacious interaction. Each system is characterized by a determinateness which disallows variance to that system as it was established. The future is not a realm of openness and contingency, but merely discloses respectively God's plan which was firmly established from the beginning of time[130], or the already-existent state of the four-dimensional manifold.

This chapter picked up Pannenberg's desire to challenge the dominance of the natural scientific view of time. And yet we saw that Pannenberg does not sufficiently distinguish between a relativistic and Newtonian conception, and thus his own critique focuses simply on the primacy of the "whole" over "part" (a tenet which the philosophy of Minkowski space-time does not reject), and fails adequately to distinguish between the underpinnings of the temporal theories presented by process philosophy and the "philosophy of the manifold". As a consequence, Pannenberg's critique is left unclear and uncertain. Is he attempting an ontological critique? If so, it quickly becomes tied up in extra-biblical issues of God's eternity and omnipresence and reintroduces an ontic distinction between geometrical and divine space—a distinction Pannenberg wishes to avoid. On the other hand, if Pannenberg's is attempting an epistemological critique, then there remain few avenues out of a Bergsonian division and segregation of knowledge—again a result that Pannenberg seeks to avoid. This is not to suggest that Pannenberg's *intention* in critiquing relativity theory or the general dominance of the natural sciences in discourses on time is misplaced. We should be wary of the unquestioned adoption of relativity theory into philosophy, and the over-eager importation of the philosophy of relativity theory into theology.

---

129 Pannenberg, "Heilsgeschehen und Geschichte I", 222; also Pannenberg, "The Biblical Understanding of Reality", 15.

130 Moltmann, *Theology of Hope*, 122.

Therefore, rather than dismiss Pannenberg's aims, this chapter sought to strengthen Pannenberg's position and closed with an extended critique of the philosophy of Minkowski space-time, highlighting the limitations of relativistic mechanics and calling into question the faith which sees it as an adequate and comprehensive theory not only of time but of the "totality of all that is". We saw that relativistic mechanics is unable satisfactorily to account for growth and development; non-substantial reality is beyond its grasp; and with this failing comes an inability to deal with such basic and central aspects of reality as information, pattern, language, pain, or consciousness.

The static four-dimensional block universe can at most describe a very limited subset of reality. Thus we can either follow the radically reductionistic path of physical scientism and declare everything outside that subset as delusional and non-existent, or push on for a complementary and more comprehensive understanding of time and reality.

CHAPTER FOUR

PANNENBERG'S CONCEPT OF TIME

4.1 Introduction

If Wolfhart Pannenberg is to stay true to his intentions of offering a theological understanding of the world with universal validity[1], then his objective theory of time will need to meet two criteria. First, it will need to address the practical insufficiencies of those temporal models offered by the natural sciences, giving a more complete understanding that opens up insights beyond what the sciences offer. In this respect, his concept would need to engage with the insufficiencies of the philosophy which underlie and support special relativity's view of time. Can an inclusive theological conception of time provide for an open future, for the existence of novelty? Can it provide a framework for change, growth and development? Can it deal with the existence of non-substantial realities? And can a theological conception of time achieve these tasks without bypassing the interdisciplinary discourse and seeking refuge in an exclusive, inner-theological "private sphere"[2], where the subject matter can be defined purely in theological terms distinct from, and inaccessible to, the natural (empirical and theoretical) sciences? In short: Can theology provide a theory of time which meets the challenges of empirical observation, but takes us further than the conceptions offered in theoretical physics?

Second, such a theory must not only be able to meet the external criterion of conformity to empirical observation of the natural world. As a *theological* conception, it must be able to meet the criterion of coherency

1 Pannenberg, *Anthropology in Theological Perspective,* 15.

2 Cf. ibid. 14.

with the broader theological traditions. In essence, a new theological theory of time must not only be able to address perceived insufficiencies in the ruling physical concepts of time, or simply account for (or prove its coherency with) empirical observation. Such a theory must do justice to the Christian tradition. The theory must be recognizably theological. Can such a unified "physico-theological" system be acceptable to rigorous scientific investigation, address the problematic triangular matrix of divine power, human action and suffering within history, and still remain recognizably Christian? To meet the requirements of a unified system both the physical and the theological aspects must be represented satisfactorily: in this interdisciplinary discussion, there is little point suggesting a new theological system if it fails to engage with or account for views of the universe held by the natural sciences; likewise, we cannot present a scientific, deterministic view of the universe if that conception cannot be made to cohere with our theological traditions. Thus Pannenberg faces a considerable interdisciplinary challenge.[3]

---

3 Unfortunately, Pannenberg has faced difficulties with his attempts to develop such universal, interdisciplinary claims before. His magisterial anthropological study, *Anthropology in Theological Perspective,* was broadly criticized for being too closely aligned with modern anthropological and psychological theory (particularly the work of G. H. Mead) at the expense of biblical and theological conceptions. Cf. e.g. F. LeRon Shults, *The Postfoundationalist Task of Theology: Wolfhart Pannenberg and the New Theological Rationality* (Grand Rapids: Eerdmans, 1999), 15. In that instance, Pannenberg's focus on an overly simplistic correspondence between *Weltoffenheit* and the *imago Dei* (*Anthropology in Theological Perspective,* 20–1, 43–79) and his narrow understanding of sin as self-referentiality (ibid. 20–1, 80–153)—while to a degree appropriate in the German context—did not satisfactory represent pluralistic societies or cohere well to the testimony of the plurality of biblical witnesses on the reality and nature of sin or the meaning of the *imago Dei* (we will return to this critique in the following chapter). On the plurality of views evident in the Old Testament alone, cf. e.g. Horst Dietrich Preuss, *Old Testament Theology,* ii, trans. Leo G. Perdue (Louisville, KT: Westminster John Knox, 1995), 170–7. Proponents of social anthropology also found the attempt at a forced amalgamation under a "universal" claim insufficient, and have accused Pannenberg of favouring theology over anthropology. This leads to a particularly ambivalent view of the entire programme where neither side is satisfied: "Pannenberg's extensive *Anthropology in Theological Perspective* is, essentially, a philosophical theology of 'doctrines of man', and

We should not underestimate the difficulty of this challenge, or Pannenberg's attempts to answer it. As we saw in the previous chapter, attempts to address the interrelation between concepts of determinism, divine power and temporality led to unavoidable problems with the concept of evil. Thus, in this respect, in order for Pannenberg's work to be judged as at least having engaged satisfactorily with the problem from a theological perspective it must either (1) provide room for both the efficacious actions of God in time (who can act in time, for example, in response to intercessory prayer) and human beings (who in turn can respond to God and God's calling) or (2) provide satisfactory theological reasons as to why the Church can either surrender belief in efficacious divine action and human free will in time, and display why the problem of evil in history is therefore irrelevant.

This chapter will begin with an examination of the impulses to Pannenberg's work (§4.2), namely his concerns about the dominance of inertia, causation and determinism in the underlying philosophy of the natural sciences, the abandonment of contingency both in the natural sciences and in the philosophy of history (embodied particularly in the analogical method) and his strong desire to defend what he perceives as the central pillars of the Christian faith (particularly, the historical resurrection of Christ) as radically novel and contingent acts which rupture the narrow epistemological preconceptions of both the historical and natural sciences.

This will then lead into a detailed examination of Pannenberg's concept of time (§4.3), as it took shape from his earliest writings, with its stress on the fundamental idea of *Rückbindung* (§4.3.1), and its dependency upon the sovereignty, power, rule, and underlying faithfulness of God. This will be followed by an investigation of Pannenberg's use of the concept of "control" to define the limits of the present (§4.3.2), and the ways in which this feeds into his view of eternity as God's particu-

---

its engagement with social anthropology is limited". He "is weak when criticizing anthropologists" and "inadequately criticizes anthropology" in general; but his appropriation of the data of nontheological anthropology (*Anthropology in Theological Perspective,* 18–19) displays how Pannenberg's "underlying method inevitably subordinates the cultural perspective to theology", Douglas Davies, *Anthropology and Theology* (Oxford: Berg, 2002), 6; cf. also Shults, *The Postfoundationalist Task of Theology,* 15–16.

lar frame of reference (§4.3.3). I will suggest that Pannenberg's work on time embodies a conception which is equally as linear and objective as Minkowski space-time, but which replaces relativity's fragmentation of referentiality towards the absolute with a unireferential view: the primacy of *God's* frame of reference for our understanding of time. In this sense, we can then classify Pannenberg's system as *linear* (having a clearly defined beginning and end), *objective* (in that the ontological "truth of time" lies outside human consciousness), and *unireferential* (taking it's primary and overarching point of reference from God and in particular God's eternity). Thus we will see that the clash that occurs between Pannenberg and the natural sciences on time is not so much based on conflicting theoretical presuppositions, since both share a view of time as objective and linear. Rather we will see that the clash arises from the confrontation of two inherently "imperialistic" systems, both of which seek to claim absolute authority and universal validity. The core difference is simply that Pannenberg's system places the unilateral and absolute power of God at the head of that system, replacing the naturally deterministic structure of the physical universe. To use a political metaphor, in the arms race to absolute power, Pannenberg seeks to overwhelm the simple deterministic "laws" of nature with an appeal to the higher and absolute rule of an omnipotent God.

4.2 Impulses to Pannenberg's Concept of Time

While Pannenberg's explicit critique of space-time theory does not occur until quite late in his work, his concern with the dominance of scientific theories of time is already evident in his earliest essays, and would become a leading motif through the course of his career. These early concerns revolved especially around what Pannenberg believed to be misdirections in the theological understanding of history. Here his focus is particularly directed at the way Bultmann and Gogarten (and their followers in the school of existential theology) have dissolved and devalued history by transforming it into a mere "historicity of existence".[4]

4 Pannenberg, "Heilsgeschehen und Geschichte 1", 218. See also Pannenberg, "Kerygma und Geschichte", 130–2, 133–4.

Likewise, he critiques Kähler, Barth and Hofmann for fracturing history, leaving us with a mundane and inconsequential husk of time and a dissociated "suprahistorical kernel" where "the real content of faith" resides.[5] Pannenberg astutely diagnoses the underlying cause of these movements as an attempt to defend "against critical-historical investigation and the scientific verification of events which seemed to leave no room for redemptive events"[6]—a process begun in earnest with David Hume's treatise on miracles.[7] For Pannenberg, it was this fear of scientific and historical processes which led existential and dialectical theology to retreat into a safe harbour beyond the reach of the scientific and historical sciences.[8]

Pannenberg is quite right to be wary of such a retreat on the part of theology. It would be devastating for theology to withdraw itself from broader academic truth claims under the presumption that it might satisfy itself with its own segregated concepts of truth and reality. "Theology must not become prey to the all too comfortable escape to the idea of creation on a special or exclusively theological level which is inaccessible to any critique by the natural sciences."[9] Pannenberg is also right to see such a move as achieving exactly the opposite of what it seeks: rather than "protection from critique by the natural sciences", such a move would simply serve to demonstrate "the irrelevancy of theological statements".[10] Yet as we saw in the previous chapter, due to underlying presuppositions regarding their respective objects of investigation, it may be impossible to maintain the independence of the natural sciences and theology without theology at least appearing to operate within an

---

5 Pannenberg, "Heilsgeschehen und Geschichte 1", 218. See also Pannenberg, "Kerygma und Geschichte", 129–30, 132–3.

6 Pannenberg, "Heilsgeschehen und Geschichte 1", 218.

7 Cf. David Hume, *An Enquiry Concerning Human Understanding,* ed. T. L. Beauchamp (Oxford: OUP, 1999), sect. x.

8 "Darum flüchtete die heilsgeschichtliche Theologie in den vermeintlich vor der historisch-kritischen Flut sicheren Hafen einer Übergeschichte, oder—mit Barth—einer Urgeschichte", Pannenberg, "Heilsgeschehen und Geschichte 1", 218.

9 Pannenberg, "Contingency and Natural Law", 74.

10 Ibid. Pannenberg rarely misses an opportunity to make this argument about the looming irrelevancy of theology before the natural sciences without referencing Barth who "has still chosen this way"; cf. ibid. 113 n. 3.

exclusive epistemological sphere. We are simply involved in examining different objects using different methodologies. While Pannenberg appreciates this difficulty and aims for "a common ground ... to which scientific and theological problematics can be referred without losing sight of the *specific difference* of the two ways of thinking"[11], natural science's own self-limitation to empirically observable substantial reality would mean that theology must necessarily form the overarching discipline (of which natural science then forms a subset). The difficulty is that either theology limits itself to the subset accessible by the natural sciences in order to maintain the discourse, or theology deals unapologetically with its broader field and so risks alienating the natural sciences. It is precisely this dilemma which spurs on the long-running science and theology discourse.

For this reason, it will be helpful to narrow down the exact substance of Pannenberg's concern in his discourse with the natural sciences. As we saw above, Pannenberg appreciates the motives of those theologians who seek out safe harbour from the historical and natural sciences precisely in order to preserve "room for redemptive events".[12] But which events does Pannenberg have in mind? Which events in the Christian gospel do the historical and natural sciences threaten? While Pannenberg sees scientific method as a threat to "Christian eschatology" and the very "idea of God itself", his major concern runs deeper. For Pannenberg, science is unable to fully appreciate the contingent and unique in history, and this failing leads to a renunciation "of the message of Jesus' resurrection". Even worse, Pannenberg worries that the sciences see such a renunciation of the possibility of resurrection as a test of "intellectual honesty".[13] It is this impulse to defend the historicity of the resurrection that influences the development of Pannenberg's concept of time.

We see this in the following four dimensions:

11 Ibid. 76.

12 Pannenberg, "Heilsgeschehen und Geschichte 1", 218: "für das Heilsgeschehen keinen Raum mehr zu lassen schien".

13 Pannenberg, "Contingency and Natural Law", 75, emphasis added.

4.2.1 Inertia as the Foundation of Natural Law

"Inertia as a principle is already at the basis of all laws of nature."[14] With this statement, Pannenberg not only introduces his most primary concern with the natural sciences, but positions himself firmly within the framework of Newtonian physics.[15] Despite the limitations of this approach, Pannenberg's wariness with regard to the concept of inertia perceptively picks up on the way it had been used by classical physics to support a deterministic view of the universe. In this respect, classical physics presented an increasingly problematic position for theology, a problem which arose out of a confluence of lawfulness (causation) and the flow of time (temporal development). The great advances achieved by Newtonian mechanics were based upon this underlying principle of the absolute lawfulness of all physical motion, connected with Newton's theory of inertia which argued that motion was self-regulating: inertial bodies would retain their states across time until acted upon by other forces, and that these interactions were predictable since they are governed by a calculable, replicable and observable set of laws.

Whereas for Ockham, "God was still indispensable in the explanation of the physical world because without God no finite reality could *persist,*" Newton's work on inertia replaced "the dependence of physical reality on God's activity of continuous creation with the idea of self-preservation".[16] Thus "Christian apologetics, having accepted the new basis of natural philosophy provided by the principle of inertia, was now left to the unfortunate strategy of looking for gaps in the continuity of natural processes if it wanted to preserve certain occasions for *divine interference* in the natural world".[17] For Pannenberg, inertia usurped the

14 Ibid. 109.

15 Critics have noted that Pannenberg's approach gets off to a false start precisely due to his "preoccupation with a rather old-fashioned concept of inertia" (Polkinghorne, *Belief in God in an Age of Science,* 82). While post-1920s theoretical physics has moved on to discuss relativistic theories, Pannenberg is seeking entry into the discussion on the basis of a surpassed Newtonian understanding of the universe.

16 Pannenberg, "Theological Questions to Scientists", 19, emphasis added.

17 Ibid. 20, emphasis added; the resurrection being the greatest example of such divine "interference".

need for God to be involved in each and every temporal event, making "decisions" regarding the status of natural events "in each case".[18] Instead of the hand of God maintaining the universe, inertia meant that objects could persist on their own without any such divine input. Rather than objects moving in space and time under the radically contingent, direct action of God, inertia "played ... a decisive role for making nature independent toward its divine origin".[19]

Thus in order to reintroduce the continuing creative and sustaining power of God in the natural world (to reinstate God's right to *divinely interfere*), a personal conception of God's involvement in physical and temporal processes needed to be reclaimed. Furthermore, if such involvement was to be truly personal, then it must also be contingent, since in such a scheme lawfulness and contingency form polar categories. It would be meaningless to speak of necessary and universal laws as "personal decisions". If theology could once more reinstate God's divine right to decide in which ways he sustains creation at each point in time, then the movement, and even very existence, of objects would once again stress the radical contingency of God's decisions in regard to temporal and historical events.[20] We shouldn't underestimate the importance Pannenberg places on the concept of decision. Whereas Newtonian mechanics appealed to the concept of inertia to provide "a natural unchangeability"[21], decision is a radically personal term expressing the motivation and involvement of a personal being[22]. *Inertia* is legalistic, whereas *decision* bears the stamp of a personal power, not of a mere structure of laws.[23]

18 Pannenberg, "Contingency and Natural Law", 109.

19 Ibid.

20 Pannenberg extends this thought one step further to its natural conclusion, that the movement of objects in time, and the cumulative events of all history, each expresses a divine decision and thus also a divine will and purpose which can reveal the will of God in history. See Wolfhart Pannenberg, "Dogmatic Theses on the Doctrine of Revelation", in id., *Revelation as History,* trans. D. Granskou (Toronto: Macmillan, 1969), 123–58, esp. 125–31.

21 Pannenberg, "Contingency and Natural Law", 109.

22 Ibid.

23 Ibid. 113.

The legalistic basis of Newtonian mechanics together with the philosophical and intellectual environment of the eighteenth century[24] promoted a temporal ontology which pictured the physical universe unfolding in strict accordance with natural law and within the framework of absolute time. The result was the replacement of a personally involved "savior God of medieval and Reformation theology" with "the clockmaker God of eighteenth-century deism".[25] The actual novelty of Newton's work lay not so much in his observations, but rather in his ability to provide standardized, laws in mathematical form which, if combined with sufficient knowledge of the objects under consideration, could thus be used for scientific *prediction:* given enough information about the initial states of a physical system, the classical physicist was confident of being able to accurately predict the future state of that system. Given complete knowledge of the past and present, science could then predict the future.[26] The central role and importance of prediction cannot be stressed enough. The vital impact of Newtonian physics comes in its ability via standardized "laws" to control and tame the contingency and uncertainty of the future.[27] It is for this reason that Newton presents such a significant marker for the beginning of modern physics. By expropriating "control of the future" from the church and placing it within the control of lawful, predictable science his work furthered the scientific revolution's move away from the exclusive authority of the church.

Within the framework of Newtonian mechanics and the epistemology of classical physics, the future no longer needed to represent an im-

---

24 Cf. R. Hahn, "Laplace and the Mechanistic Universe", in David C. Lindberg and Ronald L. Numbers (eds.), *God and Nature: Historical Essays on the Encounter Between Christianity and Science* (Berkeley: University of California Press, 1986), 256–76.

25 David C. Lindberg, "Introduction", in id. and Ronald L. Numbers (eds.), *God and Nature: Historical Essays on the Encounter Between Christianity and Science* (Berkeley: University of California Press, 1986), 1–18 at 12.

26 Cf. Eddington, *The Nature of the Physical World,* 228.

27 "What is the ground for so much ardent faith in the orthodox hypothesis that physical phenomena rest ultimately on a scheme of completely deterministic laws? ... The principal laws of Nature which have been discovered are apparently of this deterministic type, and these have furnished the great triumphs of physical prediction"; ibid. 286.

penetrable realm of uncertainty, or of contingency, but was in principle wholly predictable and knowable. Likewise, the present (as the future moment of the immediate past) was also sterilized of any contingency: the present was merely the lawful and mechanistic extension of that physical system which existed immediately prior to it. Thus classical physics was confident that with enough information it could entirely remove contingency from the physical world. Within its epistemological framework, that which still appeared to be contingent was simply the result of insufficient knowledge or calculation errors.[28] With diligence and enough data, even these final traces of uncertainty could be eliminated.[29] While in simple terms the Newtonian revolution in physics merely stressed that the past led to the present which then leads to the future, the power of the concept lay in its usefulness for understanding the universe as a self-sustaining system removed from external (divine!) contingencies.[30] Thus the impact and implications of this shift to an inertial understanding of the physical world led Pannenberg to consider the revision of the principle of inertia and its interpretation as one of the most crucial issues to be addressed in the dialogue between science and theology.[31]

---

28 The classical assumption that small errors in calculation (e.g. from the rounding of decimal places) would only ever result in minor errors in the outcome was finally dismissed by modern chaos theory; cf. Andrew Whitaker, *Einstein, Bohr and the Quantum Dilemma: From Quantum Theory to Quantum Information* (2nd edn.; Cambridge: CUP, 1996), 29.

29 "It was, of course, recognized that we were only acquainted with part of the structure of this causal scheme, but it was the settled aim of theoretical physics to discover the whole", Eddington, *The Nature of the Physical World*, 286.

30 Though it is important to note that removing the role of God in the universe was contrary to Newton's own desires. For more detail on Newton's religious aims, see either Frank Edward Manuel, *The Religion of Isaac Newton* (Oxford: Clarendon, 1974) or the discussion in Barbour, *Religion and Science*, 17–24: "Newton, however, believed that God has a continuing active role in the physical world. He said that space is eternal and immutable and is constituted by God's omnipresence. He held that gravitational attraction is not a power inherent in matter; he proposed that God may act either directly, to make bodies attract each other by exact laws, or indirectly, through an ether or a diffusion of very tenuous matter. Newton saw the laws of nature as a continuing expression of God's sovereignty and will, acting in a regular way"; ibid. 22–3.

31 Pannenberg, "Theological Questions to Scientists", 20.

4.2.2 Development and Determinism: The Hegelian Critique

Pannenberg notes that two further points arise out of this programme developed by classical physics, points closely linked to the perceived inertial underpinnings of natural law. The first is that an ordered, law-bound development (where the past leads to the present and future) provides a developmental, or *germinating* concept of temporality.[32] If the physical world and historical occurrences develop from a seed or germ-like kernel, then time and history must become simply the expression and unfolding of the pre-determined tendencies already contained within that embedded, internal seed. Pannenberg sees this especially reinforced by theologies which limit God's act of creation to "the beginning of the world".[33] Here theology is transformed into a deistic system where the universe progresses on its own under the influence of causal natural law, disconnected from the personal and contingent acts of the divine.[34] The entire process of nature "cannot be imagined as a whole as development—in any case, not as development in the strictest sense of an entelechially directed process, that is, not as the unfolding of germlike beginnings".[35] Such a development would subject creation again to the rule of necessity under natural law and rid both creation and the temporal events which compose history of their contingency, simultaneously blocking the way for the personal and contingent acts of God. "There would be no true contingency"; "Contingency and a purposefulness that directs everything from the beginning (in the sense of entelechy) exclude each

---

32 "Entfaltung einer keimhaften Anlage", Pannenberg, "Heilsgeschehen und Geschichte (Fortsetzung und Schluß)", 262f.

33 Pannenberg, "Contingency and Natural Law", 72.

34 Pannenberg draws on the statement by Otto Kirn, *Glaube und Geschichte. Eine dogmatische Untersuchung* (Fakultätsprogramm der theologischen Fakultät in Leipzig; Leipzig, 1900), 40, that: "Giebt man dem Gesetz der Causalität die Fassung, daß jede geschichtliche Erscheinung aus dem empirisch feststellbaren Gesamtzustand der Gesellschaft erklärbar sein müsse, so schließt man die Ursprünglichkeit und Neuheit von vornherein aus, welche der Christliche Glaube für die Offenbarung Gottes in Anspruch nimmt"; Pannenberg, "Heilsgeschehen und Geschichte (Fortsetzung und Schluß)", 263.

35 Pannenberg, "Contingency and Natural Law", 83.

other".[36] Such a process of internal development based on natural lawfulness would mean that historical formations and temporal events could thus only express into actuality that potential which is already contained within them. Pannenberg is right to be concerned that such a system would allow no room for anything qualitatively *new* in history, since *the new* can only ever simply be the developmental expression of that potential which is already contained in embryonic form within the object under examination. Whereas Moltmann can speak of "the vista of the future [as] already implicit, or in germ, in the past itself",[37] Pannenberg rejects such a view since it would turn the future into a determined extension of the present and past, a continuation of the conditions and structures which control the present. Therefore, Pannenberg judges this view to be "the expression of an immanent worldview, which fails to do justice to the openness of all reality on its way to the future".[38]

It is worth pausing at this point to highlight the distinctive difference here between Pannenberg and Hegel, and to stress the distance between their two systems. One of the most long-standing and persistent misunderstandings of Pannenberg's work is to view it as a form of "Christian Hegelianism".[39] Given the discussion above, it should be clear now how mistaken statements are that argue that Pannenberg shares "with Hegel ... the view that the divine goal is the entelechy of the history that God initiates and drives".[40] Yet it appears that Pannenberg will never be free from having to insist that "I am not a Hegelian"![41] Any sense of *devel-*

36 Ibid.

37 Moltmann, *God in Creation,* 131.

38 Pannenberg, "Heilsgeschehen und Geschichte (Fortsetzung und Schluß)", 262: "als Ausdruck einer immanentischen Weltanschauung, welche der Offenheit alles Wirklichen zur Zukunft hin nicht gerecht wird".

39 Cf., among others, discussions in Michael Gilbertson, *God and History in the Book of Revelation: New Testament Studies in Dialogue with Pannenberg and Moltmann* (Cambridge: CUP, 2003), 11–16; Cyril O'Regan, *The Heterodox Hegel* (Albany: SUNY Press, 1994), 128; Herbert Neie, *The Doctrine of the Atonement in the Theology of Wolfhart Pannenberg* (Berlin: Walter de Gruyter, 1978), 97–8; and esp. Shults, *The Postfoundationalist Task of Theology,* 151–3.

40 Mark C. Mattes, *The Role of Justification in Contemporary Theology* (Grand Rapids: Eerdmans, 2004), 73.

41 "I just happen to think that Hegel was one of the outstanding minds in the

opment in an evolutionary sense, any entelechy, any concept of a future that develops lawfully and organically from the past, is rejected by Pannenberg as a threat to the contingency of God's will, and as a limit on God's freedom to bring about a novel future. In the following sections we will see how this view forms the basis for Pannenberg's central concept of *backward linkage* (*Rückbindung*), his idea that events move not from past to future but rather from future to past, "and so establish repeatedly in a new way connections backward".[42] If there is to exist a dynamic, open and contingent flow of history, Pannenberg can only conceive of it occurring in the arrival of contingent events from God's future, moving then into the past. Thus Hegel's postulate of a universal Absolute gradually unfolding and realizing itself in history (from past to future) is diametrically opposed to Pannenberg's theory of *Rückbindung* (which flows from future to past).

Pannenberg's concern is that a developmental system based on inertial mechanics ultimately provides no room for divine activity within history. While Newton sought to preserve the role God played in sustaining the universe, Pannenberg is acutely aware of the danger that the spirit of classical physics presented for a theology which assumed the existence of a God active in history. A utilitarian focus upon the independence of the physical creation over against the creator (even to the extent of postulating stable natural laws which are not dependent upon new and contingent acts of the creator, either at each successive point in

---

history of modern thought, one whose work sets a high standard for us to follow. ... But very few of my ideas did I actually get from Hegel—very few"; writing in Michael Bauman, *Roundtable: Conversations with European Theologians* (Grand Rapids: Baker, 1990), 48, quoted in Mark L. Y. Chan, *Christology from Within and Ahead: Hermeneutics, Contingency, and the Quest for Transcontextual Criteria in Christology* (Leiden: Brill, 2000), 214. See Neie for a discussion on those who have influenced Pannenberg's thought; Neie, *The Doctrine of the Atonement in the Theology of Wolfhart Pannenberg*, 94–101. Though see also Wolfhart Pannenberg, "La signification du Christianisme dans la philosophie de Hegel", *Archives de Philosophie*, 33, 1970 (1971) = "The Significance of Christianity in the Philosophy of Hegel", in id., *The Idea of God and Human Freedom*, trans. R. A. Wilson (Philadelphia: Westminster, 1973), 144–77.

42 Pannenberg, "Contingency and Natural Law", 83. We will return to this point below.

time or, as Newton held, at set junctures) leads either to deism at best or at worst a rigid physical determinism absent of any divine conception.[43] Therefore, Pannenberg makes his position clear: "an interpretation of the whole of history *via a concept of development even an underlying teleology* conflicts with the contingency of individual events and is therefore not only theologically but also historically unacceptable".[44]

A developmental understanding of time limits the actualization of anything radically new and prohibits the occurrence of the contingent. The germination or unfolding of seed-like tendencies leads to a determinism of the physical world through the stable laws of causation.[45]

---

43 "'Rational religion' had been intended as a support for the essentials of Christianity, but ... was to become a substitute for them", Barbour, *Religion and Science,* 21.

44 "[A]n interpretation of the whole of history using a developmental concept or even an underling teleology conflicts with the contingency of individual events and is thus not only theologically but also historically unacceptable"; Pannenberg, "Heilsgeschehen und Geschichte (Fortsetzung und Schluß)", 262. Pannenberg also appeals to Troeltsch to support his argument, noting that contingency can only exist by refusing to allow everything in history to be determined by a "developmental idealism of history" (*entwicklungsgeschichtlichen Idealismus*); ibid. n. 8.

45 It is worth noting that human life was protected from the strict physical determination of the Newtonian system using an anthropology based upon a body–soul duality: physical matter may be determined, but the human soul (as non-material) was free of physical determination. In this way, human free will could still exist in a physically determined Newtonian universe. The modern breakdown in this dualistic anthropology led to determinism becoming more problematic for our understanding of the human person. In a monistic view of the human person, "a complete determinism of the material universe cannot be divorced from determinism of the mind" (Eddington, *The Nature of the Physical World,* 298). Once the notion of a body–soul duality is abandoned, the purely physical human being and its actions then also become subject to the same deterministic process as the rest of the universe. It may be tempting to approach this dilemma by trying to find ways to exempt the human being from this deterministic system. Eddington takes the converse approach, arguing that "if we wish to emancipate mind we must to some extent emancipate the material world also". With the advent of quantum physics, and the dismantling of classical determinism, Eddington notes that there "appears to be no longer any obstacle to

4.2.3 The Analogical Method

Reductionism in the sciences that reduces reality to predictable, inertial developments governed by strict laws of nature lead us quickly to a general physical determinism. Yet what troubled Pannenberg even more was that the appeal of deterministic scientific law had started to filter into the schools of historical studies[46], introducing a deterministic scheme into the epistemology underlying the study of historical events.

---

this emancipation"; ibid. See here Stephen Lakkis, "Reclaiming Christian Anthropology in a Post-Soul Era", *Taiwan Journal of Theology,* 37 (2014): in press.

46 Pannenberg points at Friedrich Mildenberger as a characteristic example; cf. Friedrich Mildenberger, "Kerygmatische Historie?", *Evangelische Theologie* 18 (1958): 419–24, who in turn appeals to Heinrich von Sybel in order to argue that "'absolute lawfulness in development' [is] the precondition of historical research"; Pannenberg, "Heilsgeschehen und Geschichte (Fortsetzung und Schluß)", 266–7 n. 22. This is characteristic of the eighteenth- and nineteenth-century modernist approach to historiography, which aimed at stripping away the confusing particularity of history and isolating the eternal truths embedded within, a process which could expose universally applicable historical laws (in parallel to scientific laws). Under the influence of Descartes' "clear and distinct ideas" Spinoza and Gabler sought to strip away historical particularity in order to better reveal the existence of the universal within. This is not to say that Spinoza was against the collection of detailed historical information; on the contrary, in his examination of biblical texts for example, Spinoza argued that these texts only achieve a clear meaning when examined in the light of their history and context. Likewise, in the study of historical occurrences, it was vital to understand the historical particularity which formed the context of that event. However, this process of recovering historical particularity was merely a preliminary stage. The truly valuable task of the historian came then with the sifting-out of those particularities in order to find the embedded and universal kernel of eternal truth. The investigation of historical particulars was only important because it provided the means for overcoming their transience and contingency. Cf. here esp. B. Spinoza, "Tractatus theologico-politicus", in id., *The Chief Works of Benedict de Spinoza,* trans. R. H. M. Elwes (London: George Bell, 1909), 101f. For Pannenberg, the aim of positivistic historians was to discover laws in history in parallel to the natural sciences, displaying an exaggerated interest in the *typical* at the expense of the *particular:* "A one-sided orientation to the typical threatens to flatten out the special particularity of phenomena, which must

Pannenberg levels his critique in particular at the analogical method.[47] "Analogy ... states that that which is difficult to understand, that which is relatively obscure is to be explained and judged by that which is familiar to the researcher".[48] Essentially, the historicity of events is to be judged by whether the historian has experiences of similar events. In the analogical method we see the confluence of the experience of the individual historian with the material and traditions which the historian takes to be authentic (judged by personal experiences), which then together produce a criterion for evaluating the historicity of an event and its subsequent interpretation.[49] In his critique of the principle of analogy, Pannenberg's main concern is the defence of God's freedom. It is this concern which prompts him to take issue with the concept, and particularly with Troeltsch's claim that "Correspondence with normal, ordinary,

---

be the actual interest of history. In this way, history becomes constructive"; Pannenberg, "Heilsgeschehen und Geschichte (Fortsetzung und Schluß)", 265.

47 Cf. esp. Pannenberg, "Heilsgeschehen und Geschichte (Fortsetzung und Schluß)", 263–7.

48 "Analogie ... besagt, daß das schwer Verständliche, vergleichsweise Dunkle, vom für den Forscher Näherliegenden aus aufzufassen und zu beurteilen ist"; ibid. 263.

49 This conflict with the concept of analogy can be traced back to Pannenberg's earliest works: Wolfhart Pannenberg, "Zur Bedeutung des Analogiegedankens bei Karl Barth. Eine Auseinandersetzung mit Urs von Balthasar", in *Theologische Literaturzeitung*, 78 (1953): 17–24; Wolfhart Pannenberg, "H. Wagner: Existenz, Analogie und Dialektik, München 1953", in *Theologische Literaturzeitung*, 79 (1954): 318–20; Wolfhart Pannenberg, "Analogie", in *Evangelisches Kirchenlexikon*, i. 113–14; Wolfhart Pannenberg, "H. Lyttkens: The Analogy between God and the World, Uppsala 1952", in Ernst Wolf (ed.), *Verkündigung und Forschung. Theologischer Jahresbericht 1956/57* (München, 1957/59), 136–42; and Wolfhart Pannenberg, "Analogie", in K. Galling et al. (eds.), *Religion in Geschichte und Gegenwart* i. 350–3. This topic extends on Pannenberg's own Heidelberg habilitation, which outlined this intersection between doctrines of the revelation of God and the concept of analogy: Wolfhart Pannenberg, *Analogie und Offenbarung. Eine kritische Untersuchung der Geschichte des Analogiebegriffs in der Gotteserkenntnis: Entwurf* (typed habilitation thesis: Heidelberg, 1955) = Wolfhart Pannenberg, *Analogie und Offenbarung. Eine kritische Untersuchung zur Geschichte des Analogiebegriffs in der Lehre von der Gotteserkenntnis* (Göttingen: Vandenhoeck & Ruprecht, 2007).

or commonly witnessed events and conditions, as we know them, is the marker of the probability by which the critic can either accept events as having actually happened or abandon them. The observation of analogies between between similar past events makes it possible to ascribe probability to them and to interpret the unknown aspects of the one by the known aspects of the other".[50]

Pannenberg sees two consequences arising from such an epistemological approach. The first is that historical events, if they are to be accepted as such, can hardly be contingent, unusual or extraordinary. As Troeltsch states, the modes of occurrence of events must be normal, ordinary and replicable. Pannenberg, of course, sees such a position as clashing with the central tenet of Christian faith, *viz.*, the contingent, unusual and extraordinary event of the resurrection: "the postulate of the in principle equivalence of all events [forms] ... the main argument against the historicity of the resurrection of Jesus".[51] As a result, second, this approach leads towards a very narrow egocentrism which seeks to pro-

50 "Übereinstimmung mit normalen, gewöhnlichen oder doch mehrfach bezeugten Vorgangsweisen und Zuständen, wie wir sie kennen, ist das Kennzeichen der Wahrscheinlichkeit für die Vorgänge, die die Kritik als wirklich geschehen anerkennen oder übriglassen kann. Die Beobachtung von Analogien zwischen gleichartigen Vorgängen der Vergangenheit gibt die Möglichkeit, ihnen Wahrscheinlichkeit zuzuschreiben und das Unbekannte des einen aus dem Bekannten des andern zu deuten"; Troeltsch, "Ueber Historische und dogmatische Methode in der Theologie", in id., *Gesammelte Schriften,* ii (Tübingen: J. C. B. Mohr [Paul Siebeck], 1913), 729–753 at 732; cf. Pannenberg, "Heilsgeschehen und Geschichte (Fortsetzung und Schluß)", 263.

51 "[D]as Postulat der prinzipiellen Gleichartigkeit alles Geschehens [bildet] ... das Hauptargument gegen die Historizität der Auferstehung Jesu"; Pannenberg, "Heilsgeschehen und Geschichte (Fortsetzung und Schluß)", 266 n. 22. Note again that when Pannenberg speaks of the resurrection of Jesus, he is speaking of a historically attestable empty tomb and bodily resurrection: "Bezeichnend für die Wirksamkeit dieses Vorurteils in der gegenwärtigen theologischen Atmosphäre ist die Art, wie Fr. Mildenberger die Abhandlung Hans Frh. von Campenhausens, Der Ablauf der Ostereignisse und das leere Grab, 1952, beschuldigt, Historisches und Kerygmatisches unzulässigerweise vermischt zu haben, weil Campenhausen es wagt, als *Historiker* mit der *Möglichkeit* der Auferstehung als Erklärung für das (ihm historisch wahrscheinliche) leere Grab zu rechnen"; ibid. 267 n. 22.

claim definitive knowledge of historical events based on the limitations of the historian's own current knowledge. When judging the historicity of events, the analogical method guides historians by prompting them to accept only those events which match their own personal experiences[52], or the experiences of reputable third parties. Since we have no other normal, ordinary, and repeatedly attestable experiences of resurrection, the Easter story cannot be accepted as a historical event.[53]

Yet Pannenberg rightly sees the danger here: the analogies which the historian uses arise not from a position of objective neutrality but out of an already given world of expressions, so that the historian's deliberations are based merely on his or her current experience, or on the current levels of knowledge which compose the historian's socio-academic background. "The analogies which the historian uses to understand the

---

52 A similar approach to the analogical method can be seen in the empathy (*Verstehen*) of the nineteenth-century German historicists. Dilthey, taking as his base Spinoza's and Schleiermacher's hermeneutics for biblical interpretation, argued that just as biblical verses needed to be understood in context, so too all historical situations needed to be understood within a whole; see R. E. Palmer, *Hermeneutics: Interpretation Theory in Schleiermacher, Dilthey, Heidegger, and Gadamer* (Evanston, IL: Northwestern University Press, 1969), 94, 98–123. To accomplish this task, historians must immerse themselves not only in as much historical data as possible, but to be immersed also into the place of the very object of enquiry. Only via this intuition could historians' minds penetrate to "the inwardness of events"; David Bebbington, *Patterns in History,* 108. In his autobiography, R. G. Collingwood provides a perfect example of this kind of *Verstehen* at work. For Collingwood, the historian must penetrate beyond the outside of an event to the "inside", to the thought of the person in the past, totally identifying one's mind with that person: "I plunge beneath the surface of my mind and there live a life in which I not merely think about Nelson but am Nelson"; R. G. Collingwood, *An Autobiography* (London: OUP, 1939), 112. *Verstehen,* as an analogy of feeling, was thus not an irrational activity, but the closest kind of artistic (and thus unverifiable) intuition one human being (the intuiting historian) could have with another (the historical figure).

53 David Hume's discussion *Of Miracles* provides a powerful example of using the "general experience" of the "uniformity of nature" to dismiss as improbable that which goes against our current levels of knowledge and experience, cf. David Hume, *An Enquiry Concerning Human Understanding,* ed. T. L. Beauchamp (Oxford: OUP, 1999), 169–87.

traditions as expressions of possible behaviour, always have their point of departure in a realm of expression familiar to the historian, never in value-free sensory experience".[54] Indeed, the analogical method may be used to trim away the particularity of a recorded event in order to force it to cohere to a predetermined set of acceptable events within the historian's personal range of experience. Troeltsch speaks here of "the omnipotence of analogy" (*Allmacht der Analogie*)[55] which ensures that all differences should be comprehended and subsumed into a uniform, universal homogeneity.[56] An example can be seen in the literary studies assumption that legends (be they stories of great deeds, hagiographies, folklore, etc.) often possess a historical *core* which has then been embellished by respective social groups. Since contemporary knowledge disallows acceptance of much of the particularity contained in the legends of primitive or early cultures, this particularity is trimmed away in the hope that we may then find the "solid core of authentic historical information".[57]

Pannenberg points us to a fundamental epistemological tension in historical studies: on the one hand, history is seen as the realm of the contingent and particular, yet on the other hand, the desire to find rules and similarities within history quickly sacrifices all particularity in these events: one finally assumes a fundamental similarity (*Gleichartigkeit*) of

54 "Die Analogien, mit deren Hilfe der Historiker das Überlieferte als Ausdruck möglichen Verhaltens versteht, haben ihren Ausgangspunkt also immer schon in einer Ausdruckswelt, in der der Historiker zu Hause ist, nie in einer wertfreien Sinneserfahrung"; Pannenberg, "Heilsgeschehen und Geschichte (Fortsetzung und Schluß)", 263–4.

55 A "prinzipielle Gleichartigkeit alles historischen Geschehens ... die freilich keine Gleichheit ist, sondern den Unterscheiden allen möglichen Raum läßt, im übrigen aber jedesmal einen Kern gemeinsamer Gleichartigkeit voraussetzt, von dem aus die Unterschiede begriffen und nachgefühlt werden können", Troeltsch, "Ueber Historische und dogmatische Methode in der Theologie", 732.

56 Pannenberg, "Heilsgeschehen und Geschichte (Fortsetzung und Schluß)", 264–5.

57 For an example of this methodological approach in action in religious studies, see e.g. Walker's overview of modern scholarship on the hagiography of the Persian martyrs: Joel Thomas Walker, *The Legend of Mar Qardagh: Narrative and Christian Heroism in Late Antique Iraq* (Berkeley: University of California Press, 2006), 100.

all reality based on one's own current knowledge of the world.[58] Thus if historical events are judged against the criterion presented by the analogical method, then no event in the past or future can ever be considered qualitatively different and truly novel from any event within the sociological–scientific sphere of the historian's present.[59]

At issue for Pannenberg is the ability of theologians to preserve a fundamental characteristic of the Judeo-Christian understanding of the activity of the transcendent God, "whose being cannot be expressed adequately in any cosmic order, yet who remains free over against all such orders", constantly creating something new and unprecedented.[60] If theology must focus on the particular, individual and contingent in order to discover the activity of the creative God in history, then any methodology or epistemology which seeks to rid the historical of that particularity and contingency necessarily contradicts the aims and goals of theology.[61] Thus, just as the issue of inertia represented for Pannenberg the central problem between theology and the natural sciences, the analogical method (as an extension of scientific, causal method) represents the central problem and "critical point" between historians and theologians.[62]

---

58 "Eine im weltanschaulichen Sinne anthropozentrische Verengung des historisch-kritischen Fragens ergibt sich erst dann, wenn nicht nur von Fall zu Fall konkrete Analogien aufgewiesen werden, sondern eine prinzipielle Gleichartigkeit alles Wirklichen vom jeweils nächstliegenden Umkreis der Erfahrung und Forschung aus postuliert wird"; Pannenberg, "Heilsgeschehen und Geschichte (Fortsetzung und Schluß)", 264.

59 Eddington notes that similar difficulties arise in the natural sciences when a homogenized reality is assumed based on the limits of scientific experience; Eddington, *The Nature of the Physical World,* 120–1. Like Pannenberg, Eddington's argument is not that current levels of scientific understanding should be used to *limit* scientific experience, but rather that the contingencies and particularities of both history and nature should act as the foundational data which *expands* the limits of current knowledge.

60 Pannenberg, "Heilsgeschehen und Geschichte (Fortsetzung und Schluß)", 266.

61 Ibid.

62 "Der eigentlich kritische Punkt für das Verhältnis der historischen Methode zur Theologie liegt … beim Analogieprinzip des historischen Verstehens"; ibid. 262–3.

4.2.4 Defending the Resurrection

Pannenberg's concern with the analogical method and the determinism of a developmental view of time extends beyond a general disagreement on methodological grounds, but is rather directed at a concrete case. In his early work, this is only referred to obliquely: "Analogies ... can ... hardly ... serve as a criterion of the reality of an event claimed by tradition"[63]; and if "a reported event explodes the analogy of what is otherwise usual or commonly witnessed, that is no reason in itself to dispute its facticity".[64] The footnotes directly identify the "reported event" "claimed by tradition" that concerns Pannenberg so much: "the historicity of the resurrection of Jesus".[65] While Pannenberg's view of the resurrection is developed further in various smaller texts[66], the publication of his Christology[67] in 1964 finally provides a more detailed presentation of Pannenberg's understanding of Jesus' resurrection.

Pannenberg identifies two major strands underlying the Easter traditions: "traditions about appearances of the resurrected Lord, and the traditions about the discovery of Jesus' empty grave".[68] Beginning with the "appearance" traditions, Pannenberg examines Paul's christophany as the most comprehensive account of such appearances, and notes five aspects involved in this "light phenomenon".[69] (1) There is a clear connection be-

63 Ibid. 266.

64 Ibid.

65 Ibid. 266 n. 22.

66 A good preliminary discussion of the historicity of the resurrection can be found in Pannenberg's essay: "Dogmatic Theses on the Doctrine of Revelation", 123–58, based on work from October 1960. For a later discussion, see Wolfhart Pannenberg, "The Historical Jesus as a Challenge to Christology", *Dialog*, 37/1 (Winter 1998): 22–7.

67 Wolfhart Pannenberg, *Grundzüge der Christologie* (Gütersloh: Gütersloher Verlag, 1964) = *Jesus—God and Man*, trans. L. L. Wilkins and D. A. Priebe (Philadelphia: Westminster, 1968). References to this text are taken from the 1977 English second edition.

68 Pannenberg, *Jesus—God and Man*, 88.

69 For the following, cf. ibid. 92f. It is interesting to note the similarity here between what Pannenberg describes as the "light phenomenon" which appeared to Paul, and his own "light experience" which marked his own teenage

tween the light experience and the person of Jesus of Nazareth. "God has revealed his Son to him (Gal 1:16); Paul has seen the Lord Jesus Christ (1 Cor 9:1)". (2) Paul did not see a physical body but rather a "spiritual" one, a σῶμα πνευματικόν. (3) The appearance did not take place on earth, but occurred from "heaven". (4) It is highly likely that the appearance to Paul on the road to Damascus "could have happened as a light phenomenon". And (5) the appearance to Paul contained an auditory aspect. The majority of these features, according to Pannenberg, were contained in other reported Easter appearances of the risen Jesus. Furthermore, in addition to the very nature of the event as described, since this appearance to Paul was "not an event that was visible to everyone ... an event of this sort must be designated as a vision".[70]

At this point, Pannenberg makes two important moves. The first is to deny that Paul's vision of the risen Jesus is in some way a-historical or "imaginary" merely because of its status as a vision.[71] "The term 'vision' can only express something about the subjective mode of experience [of an event], not something about the reality of an event experienced in this form".[72] The second move is to link the reality of these visions with the empty tomb traditions, which Pannenberg sees as providing the underlying confirmation for those resurrection appearances.[73]

---

conversion; cf. Pannenberg, "God's Presence in History", 261.

70 "If someone sees something that others present are not able to see, then it involves a vision. Grass correctly emphasizes that the form of experience in which the appearances of the resurrected Lord were perceived can only be designated by the term 'vision'.... It involves extraordinary sights that were not imparted to all and also (in any event in the case of Paul) were not perceived by all present. These are the characteristics of visions"; ibid. 93.

71 Pannenberg argues that this is not to suggest that all events described by the early church as visions possessed the same levels of historicity. The early church was familiar with a range of visionary experiences. However, "one must at the same time take into consideration that primitive Christianity itself apparently knew how to distinguish between ecstatic visionary experiences and the fundamental encounters with the resurrected Lord"; ibid. 94.

72 Ibid. 95.

73 For a detailed summation of this argument from *Jesus—God and Man,* 88–106, see Pannenberg, *Systematic Theology,* ii. 343–63.

> Decisive for confidence in the facticity of the resurrection of Jesus as the Christian message proclaims it are the primitive Christian testimonies to the appearance of the risen Lord to his disciples, along with the discovery of the empty tomb of Jesus in Jerusalem. We cannot accept these testimonies blindly on mere authority ... but must see whether they hold up to the kind of testing to which we submit, and by which we prove, other reported facts.[74]

At this point, Pannenberg's concentration on the appearances of the risen Jesus wanes and is replaced by an amplified focus on the empty tomb and the bodily resurrection of Jesus as the most significant kernel of truth underlying the resurrection traditions.

> Those who want to dispute the empty tomb of Jesus must show that contemporary Jewish witnesses to belief in the resurrection included some who did not think the resurrection of the dead need have anything at all to do with the body in the tomb. They must also assume that such views (not thus far attested) were sufficiently popular in Palestine, else the first Christians could not have successfully preached the resurrection of Jesus if his body had been intact in the tomb. Even then, an explanation would still be needed as to why the Gospels present debates between Christians and their opponents regarding what happened to the body. So long as no proofs are offered on these points, we must assume that the tomb of Jesus was in fact empty.[75]

In so doing, we note the convergence of several points of argument. (1) Visions of the risen Jesus are "real", in the sense that they represent a historical reality and are not merely imaginative or "suprahistorical". (2) The appearances find their basis in the Easter tradition of the empty tomb. (3) The assertion that Jesus is raised from the dead, and that the tomb is empty, are to be understood as assertions regarding the facticity of a historical event.[76] And (4) as historical events, these reports must be

---

74 Pannenberg, *Systematic Theology,* ii. 352–3.

75 Ibid. 358–9.

76 "Something happened in which the disciples in these [resurrection] appearances were confronted with a reality which also in our language cannot be expressed in any other way than by the symbolic and metaphorical expression of the hope beyond death, the resurrection from the dead. Please understand me

examined in line with other historical events, namely, they must be open to historical investigation using standard historical methodology, in line with the historical investigation of any other reported event.

> We cannot separate the question whether the resurrection is part of the basis of faith from that of the facticity or historicity of the event. ... If the resurrection was not a historical event but was still the decisive starting point of primitive Christian proclamation, as Bultmann assumed, then we can ascribe no significance at all to historical knowledge as the basis of our confession of Christ. If, however, we must go back to the historical Jesus to protect the kerygma against the suspicion that it is a mere myth, then, assuming that the resurrection is not historical, we must seek agreement between the kerygma and the message of Jesus without including the subject of Easter. In this case ... the development of the christological tradition would be unintelligible as a historical process. But if we can establish the facticity of the Easter event (in a way that has yet to be described), then it is possible that the history of the confession of Christ up to the formation of the christological and trinitarian dogma of the church can be viewed as an explication of the meaning that is intrinsic to the history of Jesus in the light of the Easter event.[77]

With this move, Pannenberg closes his epistemological circle and returns us to the methodological problems faced by modern historiography: namely, the attribution of historical facticity based upon the analogical method. In short, if the reality of the *kerygma* is based upon the historicity of the resurrection, and if the experiences of the resurrection are based upon the historical facticity of the empty tomb, then our point of access to that historical fact must be through the study of history (and not through a Bultmannian or Barthian appeal to a psychological or supra-historical reality for the resurrection. For Pannenberg, if the study of history is dependent upon flawed reasoning and a flawed methodo-

---

correctly: Only the name we give to this event is symbolic, metaphorical, but not the reality of the event itself. The latter is so absolutely unique that we have no other name for this than ... metaphorical expression. In this sense, the resurrection of Jesus is an historical event, an event that really happened at that time"; Pannenberg, "Did Jesus Really Rise From the Dead?", 135; cf. also Pannenberg, "History and the Reality of the Resurrection", 62–72.

77 Pannenberg, *Systematic Theology*, ii. 285.

logical approach via the analogical method (which presupposes that the dead do not rise), then it is the task of theology to describe and approach that facticity in a new way.

4.2.5 Summary: Rule by Natural Laws

Despite Newton's intentions, the strict lawfulness of classical physics quickly usurped the need for divine intervention within the natural world. Newtonian physics displayed the self-sufficiency and regularity of a universe founded on inertial laws, and the substantial powers of predictability that such regularity and lawfulness provided. Furthermore, classical physics believed that given complete knowledge about the past and present, it should have no difficulty in predicting the future. The universe progressed in orderly lawful ways and displayed a natural development where the configurations of the past were then expressed and developed in the future. This development effectively removed novelty from the physical system, and contingency was recast as something inherently knowable. The future was no longer conceived as qualitatively different from the present or past: those universal laws which apply in the present also have jurisdiction over the future, just as they controlled the past. Likewise, that which is possible (according to those laws) in the present, is equally possible in each time frame; and conversely, that which the physical laws declare to be impossible in the present remain equally impossible in all time frames. In this respect, the great accomplishment of classical science was its ability to replace the cult as that social institution which could tame the contingency, novelty, and frightening uncertainty of the future. The cult sought to control the uncertainty of the future by drawing on the powers of the gods. In contrast, classical physics addressed the uncertainty of the future by using natural law not only to predict the future but to eliminate the concept of contingency itself. That which is naturally possible in the future is that which can be extrapolated in accordance with stable physical laws from that state of the physical world temporally prior to it, and with those events which are analogous to our current experience. In accordance with the lawful rules of development, a future event is merely the expression of that state of affairs which immediately preceded it. The future, as the flowering of the past and present, can contain nothing radically new, indeed nothing that was not already contained in the present and past. If this approach is paired with a physical reductionism, then what we are left with is a strict determinism not only of the natural world, but even of the human person. The deterministic laws of nature direct each and every natural and temporal event, determining and specifying the ways in which they occur and appear as the present slowly develops into the future. This confluence of determinism, physical reductionism and the rejection of novelty is the basis for Pannenberg's heavy critique of any system of development based on set, natural laws. Such an approach is theologically and historically unacceptable.[78]

Equally disturbing for Pannenberg is the parallel tendency among historiographers to search for homogenizing laws of history, expressed in the analogical method which assumes a fundamental similarity to all reality judged according to one's own current knowledge of the world.[79] The analogical method, in its work to strip away all particularity from historical events in order to isolate an underlying, analogous truth, validates and reinforces the current intellectual paradigm. Pannenberg's major concern with this approach, in relation to Christian faith in the resurrection of Jesus, is that it already begins with the presupposition that the dead do not rise. Thus theology is in the impossible position of having to defend the bodily resurrection of Jesus using a historical philosophy that rejects *a priori* the possibility of resurrection and historical novelty.[80] "The possibilities that [the historian] can consider ... will depend upon the understanding of reality that he brings with him to the task. If the historian approaches his work with the conviction that 'the dead do not

78 Pannenberg, "Heilsgeschehen und Geschichte 1", 262.

79 Ibid. 264.

80 "Does not the vital interest of the historian already lie at the basis of all historical investigation, even though such an interest certainly cannot be permitted to prejudice the results of the inquiry? If one pays proper attention to these presuppositions of historical inquiry itself which have been discussed for decades under the catchword 'preunderstanding' (*Vorverständnis*), one will hardly be able to call into question Paul's intention of giving a convincing historical proof by the standards of that time, in any case, not from the perspective of a historical inquiry that is supposedly disinterested"; Pannenberg, *Jesus—God and Man*, 89.

rise', then it has already been decided that Jesus also has not risen".[81] If the particularity of the resurrection is rejected from the start, then the current historical paradigm can never be opened up enough for theology to argue for the bodily resurrection of Jesus as a historical act, or to allow the epistemological horizons of historiographers and natural scientists to be broadened enough to allow the submission of that historical data which testifies to a resurrection.[82] "The possibility of the historicity of Jesus' resurrection has been opposed on the grounds that the resurrection of a dead person even in the sense of the resurrection to imperishable life would be an event that violates the laws of nature. Therefore, resurrection as a historical event is [understood to be] impossible".[83]

Finally, despite these difficulties, Pannenberg is unwilling to remove the resurrection from the field of the historical as, he argues, Bultmann and Barth have done.[84] Thus the major impetus to Pannenberg's work on time and history is to create the necessary conditions of contingency required for "divine interference"[85] in nature—to provide space for a bodily resurrection of Christ. It could rightly be argued that such an interpretation of the resurrection is not representative of the broader Christian traditions. Christian doctrine has developed a series of models for understanding the resurrection, not just a "liberal" approach (Bultmann and even John Knox), or "evangelical" approach (Barth, Künneth

81 Ibid. 97.

82 This was the point behind Thomas Kuhn's argument that scientific advancement often does not occur in an evolutionary fashion but is rather instigated by a paradigm shift, an "intellectually violent revolution" where "one conceptual world view is replaced by another"; Thomas Kuhn, *The Structure of Scientific Revolution* (Chicago: University of Chicago Press, 1996), 10ff. However, Kuhn's paradigm shift is not simply to any new paradigm but rather often a *return* to an older form which is then reconsidered and re-evaluated in new light. Thus Pannenberg rightly critiques that mentality which seeks to exclude in advance the data which would provide the basis for this type of paradigm shift in our understanding of the human being and the possibility of resurrection.

83 Pannenberg, *Jesus—God and Man*, 98.

84 Cf. Pannenberg, "Heilsgeschehen und Geschichte 1", 218, and Pannenberg, "Zur Bedeutung des Analogiegedankens bei Karl Barth. Eine Auseinandersetzung mit Urs von Balthasar", 17–24.

85 Pannenberg, "Theological Questions to Scientists", 20.

and Schillebeeckx), but also a "liberation" approach (Moltmann and Sobrino).[86] None of these interpretative models would require such a degree of radical historical and metaphysical reorganisation. Yet Pannenberg suspects that all these alternatives are reactions to determining scientific and historical models which precisely block theology from an understanding of Christ's bodily resurrection. Thus rather than pursue the polyvalent character of the resurrection event, Pannenberg instead aims to stress the absolute inner-temporality of "the event [which] took place in this world, namely, in the tomb of Jesus in Jerusalem before the visit of the women on the Sunday morning after his death".[87] Faith in Jesus Christ, he holds, should not be relegated to some field of the psychological, existential or supernatural where critical analysis cannot reach. It must be firmly grounded in history, in particular the historical resurrection of Jesus. Faith presupposes historical facts which are open to reason and rational historical enquiry and is not dependent merely upon the authoritarian statement of doctrine. "Whether or not a particular event happened two thousand years ago is not made certain by faith but only by historical research".[88]

Thorwald Lorenzen, describing Pannenberg's view on the resurrection, writes that "Theology and proclamation have withdrawn from the marketplace of human life and escaped into a sectarian existence. They have failed to assume a responsible place in modern society, and therefore they cannot fulfil their mission of giving a credible account for their faith in the categories of the modern world".[89] It is this point that Pannenberg seeks to address in his desire to defend the historical reality of the bodily resurrection. He does so by directly confronting the scientific and historical understandings of time, and offering an alternative in their place. It is to this alternative that we now turn.

---

86 I am following here the classification suggested by Thorwald Lorenzen, *Resurrection and Discipleship: Interpretive Models, Biblical Reflections, Theological Consequences* (Maryknoll, NY: Orbis, 1995).

87 Pannenberg, *Systematic Theology*, ii. 360.

88 Pannenberg, *Jesus—God and Man*, 99; "There is no justification for affirming Jesus' resurrection as an event that really happened, if it is not to be affirmed as a historical event as such".

89 Lorenzen, *Resurrection and Discipleship*, 18.

4.3 Pannenberg's Concept of Time

The point at which Pannenberg seeks to engage with the temporal and historical sciences is now clear. On the one hand, Pannenberg wishes to retain the strict historicity of the acts of God including, above all, the bodily resurrection as an event that occurred within this current flow of experienced history. Yet on the other hand, Pannenberg's appeal to that historical event is barred by what he understands to be the collusion between physical laws of nature based on deterministic, inertial, and causal processes of development (where the present and future are strictly limited by the past, and thus segregated and separated from any future contingent action by God) and the expression of these laws in the historical analogical method, which blocks all access to an event so peculiar that it appears to break natural laws and possess no analogy.

For Pannenberg, causal and lawful natural development from past to future, and the contingent freedom of the divine to introduce a novel future are mutually exclusive.[90] Since Pannenberg is unwilling to surrender the historicity of divine action, and since he is also closed to the polyvalent nature of the resurrection as anything other than a strictly historical event, his response must be to disassemble the scientific concept of deterministic natural law and the concept of time upon which they are based. This is precisely what he attempts. This section will examine the way in which Pannenberg does this, by focussing on the primacy and absolute power of God.

4.3.1 Rückbindung

Pannenberg begins with a simple observation: *If* history and the cosmos run within their own deterministic frameworks, where the past determines the present and future in turn, then there is little room left for theology to speak of the freedom of a personal God who acts in history. If we "replace the contingency of events with deterministic models of re-

90 "Contingency and a purposefulness that directs everything from the beginning (in the sense of entelechy) exclude each other"; Pannenberg, "Contingency and Natural Law", 83.

ality, the notion of a personal power behind those events is untenable".[91] And *if* the course of history and the course of the cosmos are already laid out within their own *developmental* frameworks so that the past leads to the future "as the unfolding of germlike beginnings",[92] then in such a situation all we have left is to argue about the mechanisms of progress, whether they are Hegelian, Newtonian, Darwinian, etc. But neither God nor humanity can ever escape from the underlying system of determination. Finally, *if* time develops from the past to the future with such an inertial "notion of self-movement", then the resultant entelechy can never be the grounds for anticipation of that which is "not yet attained" but merely a qualitative continuation of the same: the future simply unfolds itself out of an "already present germ" and thus can never be the basis of something "really new".[93] For these reasons, Pannenberg poses a simple thesis in response: *it cannot be the past which determines the present and the future; rather it is the future which must determine the present and the past.*[94] To support this thesis, Pannenberg develops a notion of time which runs not from past to future but rather from future to past. The motor driving this new conception of time is Pannenberg's notion of a divinely powered "backward linkage" (*Rückbindung*).[95]

Pannenberg reverses the normal understanding of time and argues that time moves from future to past. If we understand the events which compose time as moving towards us from the future, then each present moment of time, and each present event occurs within a God-created space of absolute contingency. To put it another way, the event itself can in no way develop or be predetermined by any pre-existing situation since it has no origin in past or present temporality. Its origins lie solely

91 Pannenberg, "Theology and the Kingdom of God", 58; "The basic idea of divine personality seems to be related to man's experience of the contingency of events"; "only contingent events can be conceived as personal acts".

92 Pannenberg, "Contingency and Natural Law", 83.

93 Wolfhart Pannenberg, "Appearance as the Arrival of the Future", in id., *Theology and the Kingdom of God* (Philadelphia: Westminster Press, 1977), 127–143 at 139.

94 Pannenberg, "Theology and the Kingdom of God", 59.

95 Cf. Pannenberg, "Heilsgeschehen und Geschichte (Fortsetzung und Schluß)", 285.

in God's transcendent, free will.[96] Pannenberg explains this move by arguing that if "the connection of occurrences is grounded in the faithfulness of the free God, then we do not have a continuity of persistence from the past into the future, in the sense of a development, but rather events which are in themselves contingent are, so to speak, connected backwards and related to what has been".[97] In such a system, there simply *is* no predetermining past which would place limits on God's power or on God's will to act in creation. Pannenberg encourages us then to stop thinking of each present moment as somehow shaped by the immediate past under the static laws of nature, and to see it rather as a space open to the free will of God. "Only in this way, as a backward reaching incorporation of the contingently new into what has been, but not the reverse, as a predetermining mastery and its effect, can the primary connection of history be conceived without losing its contingency".[98] Each present moment is a part of God's future which God then brings into being. God *attaches* this "moment of contingency" to the course of time so that Pannenberg can say, "The future lets go of itself to bring into being our present".[99] Thus rather than being the basis upon which all future events grow and develop, the past is merely the remnant of what were once God's multitudinous free and contingent futures, the brittle residue of countless new and contingent instances flooding into the present, having been brought directly into existence by the almighty hand of God.

The concept is both compact and complicated in its merger of several different theological and scientific aspects, five of which are vital here.

96 Ibid. 284–5.

97 "[Wenn d]er Zusammenhang der Begebenheiten in der Treue des freien Gottes gründet, dann haben wir nicht eine Kontinuität des Durchhaltens aus der Vergangenheit in die Zukunft hinein ins Auge zu fassen, etwa in der Weise einer Entwicklung, sondern die in sich kontingenten Ereignisse werden gleichsam nach rückwärts angeknüpft, auf das Gewesene zurückbezogen"; ibid. 285.

98 "Nur in dieser Weise, als rückgreifende Eingliederung des kontingent Neuen in das Gewesene, nicht aber umgekehrt als vorgreifende Lenkung und Wirkung kann der primäre Zusammenhang der Geschichte gedacht werden, ohne daß ihre Kontingenz verlorengeht"; ibid. = Pannenberg, "Redemptive Event and History", 76.

99 Pannenberg, "Theology and the Kingdom of God", 59.

4.3.1.1 *Sterile Linearity*

The theory of time Pannenberg proposes is thoroughly linear; it merely runs in the opposite direction to more common understandings of time. However, there are of course significant differences which distinguish Pannenberg's concept from the linearity of Minkowski space-time, or even from the simple temporal linearity of classical Newtonian physics.

Most obviously, as we have already seen, is that time for Pannenberg runs opposite to what the natural sciences assume. Time is pushed upon us from the future and then fades into the past. Thus the future is not an empty category waiting to be filled or populated by a growing past, nor is it a field of probabilistic chances waiting to be actualized.[100] Rather the future is the domain of God's continuing creation. In this respect, Pannenberg stresses again that "creation" is not "the beginning of the world".[101] To speak this way, he suggests, is relegate God's creative activity solely to a dark and distant past, and to leave theology open to the dangers of deism. It is the attempts to avoid such a deistic tendency that lead theology then to the second order concept of providence, which establishes (and must then deal with) a distinction between the appearance of an original divine creation and its subsequent control or direction.[102] Creation, especially *ex nihilo,* need not presuppose anything, and thus places no constraints or limits on God's activity or sovereignty over against his creation.

As the domain of God's free and unlimited creation, the future sees the generation of new, contingent and individual events. "In every event the infinite future separates itself from the finite events which until then had been hidden in this future but are now released into existence".[103] Thus we have here a discrete and atomistic view of time with two immediate consequences. First, time is once again linked to the occurrence of new

100 Ibid.

101 Pannenberg, "Contingency and Natural Law", 72.

102 "'Providence' ... [as] the divine government of the world", "is so closely related with an understanding of God in the sense of a highest consciousness which has become questionable today that it appears therefore unsuitable. Besides, it always presupposes, in distinction from the concept of creation, the existence of a world that is to be governed"; ibid. 113 n. 1.

103 Pannenberg, "Theology and the Kingdom of God", 59.

states or events and pictured as dynamic, as *flowing* or moving (though Pannenberg fails to address the problem this raises of a "meta-time" in which the "standard" time of God's discrete future moments "flow"[104]). Second, rather than the four-dimensional manifold of Minkowski space-time which extends as a static and invariable object from the absolute past into the absolute future (and thus requires no "meta-time"), Pannenberg presents us, so to speak, with a system of temporal ovulation: each event, each moment of time, is an atomistic unit released into existence from the freedom of God's infinite future to settle at the end of the temporal chain as our current "present" moment. Yet unlike true ovulation, nothing living can develop out of that which is here released. Pannenberg explicitly rejects the germlike "conception of a forward expanding development".[105] Thus each event released from God's future is thoroughly sterile and awaits the next divinely-released, temporal event to replace it as the new "present".

4.3.1.2 *The Sovereignty and Rule of God*

In order to protect the contingency of all historical acts against the determinism of natural law, Pannenberg stresses the free power and sovereignty of God. Unlike determined natural processes, God and God's future are in no way to be seen as limited. Nature can provide no limit upon the freedom of its creator, and the pure contingency of the future, the contingency of each and every moment of time, reflects the supreme freedom and authority of God to bring about something radically new.[106] Indeed for Pannenberg, absolute power and sovereignty are the defining marks of a Christian understanding of God. He states unequivocally, that if our image of God "does not comprehend the power dominating everything, [it] is therefore not a true concept of God".[107] Pannenberg is

---

104 Cf. the discussion of meta-times in the previous chapter and in Begbie, *Theology, Music and Time,* 65 n. 93.

105 Pannenberg, "Contingency and Natural Law", 106.

106 The "Christian doctrine of creation ascribes to God the bringing forth of something new. Here the contingency of the occurrence was positively understood as expressive of the freedom of the creator"; Pannenberg, "Appearance as the Arrival of the Future", 139.

107 Pannenberg, "Theology and the Kingdom of God", 60.

consistent in his repetition of this claim: "Only the God who proves himself master over all is true".[108] According to Pannenberg, if Christianity is to understand God and God's contingent acts in history then it must recognise that "The deity of God is his rule".[109] There can be no aspect of creation, be it a temporal past, a physical world, a natural law, or even a human act, which sets limits on and restricts the absolute power and freedom of God to act contingently. "God's acting is *not conditioned by anything given.* Divine acting is therefore accidental, accidental also in *the further process of occurrences in relation to that which already exists*".[110] It is worth pausing here to examine exactly what Pannenberg means with this statement. In his earlier work, Pannenberg made some minor efforts to reserve at least the concept of a human role in the course of history, even though these efforts were undeveloped and stood at odds with the body of his argument.[111] But here there is no such hesitance. Not only is it God who *acts* (i.e. as *primary* actor, as the introducer of activity and events into history) and who does so in a way *unconditioned* by existent circumstances (which are merely the residue of God's previous free and contingent divine acts), but neither does God *react* in "further processes", in *response* to that which already exists. Each individual act of God is completely and radically sovereign, unconditioned and undirected by any external reality.

108 Ibid. 55.

109 Ibid.

110 Pannenberg, "Contingency and Natural Law", 103–4, emphasis added.

111 See for example his "Heilsgeschehen und Geschichte (Fortsetzung und Schluß)" where immediately after introducing his concept of *Rückbindung* under the hand of the "free God", he adds a footnote that "Diese Rückbindung wird in der Menschengeschichte nicht als Verhältnis geistloser Fakten, sondern unter Beteiligung des Menschen Ereignis"; 285 n. 68. This leads into a long, discursive note on joint human responsibility in history, but without clarifying how events as the absolutely free and contingent acts of God can coexist with human acts in history. At least in the footnotes, Pannenberg seems interested in maintaining a motif of efficacious human action in time, but is unable to introduce it well into the body of his system. As I will discuss in the following chapter, this tension then comes to a head in Pannenberg's *Anthropology in Theological Perspective.*

The connections of the occurrences which form time and history are "established *by God* in *his* acting",[112] disconnected from any sense of efficacious human action or the natural persistence of the universe. From the perspective of human action, this is a passive understanding of time and history, and as such also provides the basis for a "passive eschatology".[113] "*Each event* is something that *happens to us* from the future which is until then still unrealized. It establishes anew the connection with that which happened earlier".[114] We are merely passive recipients open to events which encounter us from the external, divine source.

4.3.1.3 *Objectivity*

In Minkowski space-time, time formed one of the dimensions of absolute invariant reality. Thus while we could relate to time only from the standpoint of particular frames of reference within it, time *per se* was part of an objective reality beyond any constitutive influence of human consciousness. Similarly, in line with his thesis that each of the events which compose time impact upon us from the free and contingent future of God, we can also place Pannenberg within the school of objective time.

This is not to suggest that Pannenberg fails to appreciate the subjective or experiential aspects of time. In a statement that could easily have come from Eddington or any number of the supporters of a philosophy of the block universe, Pannenberg notes that "Human beings never live only in the now".[115] However, he is not speaking here *ontologically* about the temporal extension of matter into the absolute past and absolute future; the future does not yet exist for Pannenberg, since it still waits on the contingent act of God for its creation. Rather human beings merely "experience their present as heirs of a past and as its active change. They anticipate the future in fear, hope, and planning; and in the light of such anticipation of the future they return to their present and the heritage of their past".[116] Thus Pannenberg expresses a strictly psychological disposition towards what is otherwise objective time, rather than any deeper on-

112 Pannenberg, "Contingency and Natural Law", 85, emphasis added.
113 Cf. Berdyaev, *Slavery and Freedom*, 264.
114 Pannenberg, "Contingency and Natural Law", 85, emphasis added.
115 Ibid. 84.
116 Ibid.

tological connection (the ontic nature of time and its connectivity being based strictly upon *Rückbindung*[117]). Pannenberg is careful to clarify this relationship between the underlying ontic connectivity of time and our human experience of it, so that our human experience or consciousness of the unity and connectivity of time should not be mistaken for that which actually creates time's unity: "Our experience ... already *presupposes*" the unity of experienced occurrences, it does not create it.[118]

Yet this leaves Pannenberg with a potential difficulty. If events come from God's open future and are expressions of God's free and contingent will, how then do we witness a unity and coherency among these purely accidental, contingent events? And how can this coherency be see solid and predictable that the natural sciences can mistake it as an expression of natural lawfulness? In order to overcome this problem, Pannenberg postulates a unique combination between the concepts of contingency, continuity and God's faithfulness.

4.3.1.4 *Contingency*

As we have seen, Pannenberg's concept of time is an objective linear system which flows from the future to the past. The events which compose this flow of time are radically contingent in that they come into existence from the open and infinite future of God's free will. Each event is released into the stream of time as our new present moment, and in so doing pushes each previous "present" (i.e. each previous free and contingent moment brought into existence by God) into the "past". "Thus the present is not independent from [God's] future", and in this way "we see the present as an effect of the future, in contrast to the conventional assumption that past and present are the cause of the future".[119] It is this combination of the generation of events as atomistic moments from the future together with the absolute and free power of God—who, for Pannenberg, must prove himself "master over all"[120] including all time, in order to be truly considered God—which thus ensures the radical contin-

117 Ibid.: "The fact that events ... throw light back on former events is closely connected with the human experience of time".

118 Ibid. 85.

119 Pannenberg, "Theology and the Kingdom of God", 54.

120 Ibid.

gency of all events. "God ... through the transcendence of his freedom [is] the source of the contingent in the world".[121] God is the origin of all discrete events (literally all temporal events), and this necessarily then includes those events which the Christian tradition would see as unique and without parallel (such as the resurrection of Christ).

This is the point where all so-called "laws" of science thus fail. For Pannenberg, since all scientific laws as mere generalizations based on past occurrences, they can bear no relevance to the contingent, future events which God brings about. God, "as the origin of all reality does not violate scientific *descriptions* of natural processes"[122] since "*every event* in which the future becomes finitely present must be understood as a contingent act of God who places that finite reality into being".[123] Thus the task of science becomes merely *descriptive:* describing those finite realities which God has released into existence. For Pannenberg, the sciences have operated with a false epistemology, in that they assume the existence of stable physical structures and natural "laws" which regulate them and can be used for prediction. Let us look at a practical example: A chemist may argue that the burning magnesium in oxygen to produce magnesium oxide makes certain assumptions about the nature of the physical world, namely that magnesium and oxygen have stable properties which react together in lawful and therefore predictable ways to produce an expected product. In response, it is not that Pannenberg rejects this stability of the physical world, but whereas a chemist may assume and speak of physical and chemical laws that determine and predict reactions and relationships between elements, Pannenberg sees only secondary scientific *descriptions* of what are otherwise free and contingent acts of God. To speak here of natural *causality* or *lawfulness* in regard to the behaviour of elements represents for Pannenberg an epistemological and methodological confusion on the part of the sciences, since they operate under the assumption of a *causal development* from the past to the future. "Ideas of causality and the physical laws presuppose that something exists and that events do happen", yet it is "only the fact that each new event has

121 Pannenberg, "Heilsgeschehen und Geschichte (Fortsetzung und Schluß)", 284.

122 Pannenberg, "Theology and the Kingdom of God", 67, emphasis added.

123 Ibid. 60, emphasis added.

to relate itself to the existing world [that] makes it possible to describe phenomena in terms of causality. ... The combined action of innumerable events conforming themselves to the world they enter produces the conformity of processes that can be described in terms of causality and physical law".[124]

To be rejected here is also any idea that events are actively relating and conforming *themselves* to past "presents". Such a move would imply an underlying ontic capability or predisposition inherent in matter. First, this might reopen the way for a probabilistic view of the future (that the future is a field of set possibilities, of which only some will actualize[125], for example only those chemicals or compounds which possess the required valencies). Second, such an assumption would place limits on the absolute, free and contingent will of the creator to bring about new events. Thus Pannenberg is careful to clarify this point: "Every event in which the future becomes finitely present [i.e. all temporal events] must be understood as a contingent act of God", and importantly it is "God, who places that finite reality into being".[126] It is through the process of *Rückbindung*, of connecting each of these atomistic moments with the current present, that God establishes their continuity: "the continuity of nature is no longer understood as the irresistible dynamic of the already existing pushing forward, but as the building of bridges to the past that save the past from getting lost".[127] In this salvific act of faithfulness, God builds bridges for his newly created present moment, connecting them to the past.

4.3.1.5 *Faithfulness*

For Pannenberg, there can be no limitation on God's freedom or God's future. If God is the all-determining reality, if "God's being is his rule",[128] then to reduce the absolute rule of God is to attack divine sovereignty, the divine nature, and the divine's identity as "true God". "A God who would not be the origin and perfecter of ... nature could not be the power that determines all reality of being, and therefore not truly God. If

124 Ibid.
125 Cf. ibid. 59.
126 Ibid. 61.
127 Ibid. 67.
128 Pannenberg, "Theology and the Kingdom of God", 55.

theology wants to think of the deity of God, then it has to think of God as the power that determines not only human history but also nature".[129] No other determinant—either the efficacious acts of human beings in history or the lawful progression of nature—would be acceptable here for Pannenberg if Christian theology is to speak meaningfully of the deity of God. The "God of the Christian faith" must be "lord of the world, as the reality that determines everything".[130]

This position results in several important consequences for Pannenberg's understanding of time and particularly with regard to the apparent continuity of what he argues are thoroughly contingent events in time. Common experience tells us that we witness continuity between two temporal events, between one moment of time and the next, between our immediate past and our current present. Indeed, the natural sciences are founded upon this presupposition of temporal (and lawful) continuity. Yet as we have seen, Pannenberg's concept of *Rückbindung,* his claim that time flows from the future into the past under the direct guidance of the free will and powerful guidance of God, not only critiques such lawful and developmental continuity but purposefully abandons it. Where then does this perceived continuity come from? Is it purely an illusion?

Pannenberg responds by returning again to the absolute and free will of God. The lack of any inherent causality in the cosmos can only mean that the appearance of continuity in creation arises because *God wills such continuity to exist:*

> If ... the connection of occurrences is grounded in the faithfulness of the free God, then we do not have a continuity of persistence from the past into the future, in the sense of a development, but rather events which are in themselves contingent are, so to speak, connected backwards and related to what has been. It is through such backward linkage [*Rückbindung*] that continuity is repeatedly produced. *In this way is the faithfulness of God expressed. Only as a backward-reaching incorporation of the contingently new into that which has been,* and not the reverse, namely a forward-pushing direction and action, can we conceive of the primary connectedness of history, without losing it contingency.[131]

129 Pannenberg, "Contingency and Natural Law", 75.

130 Ibid. 114 n. 5.

131 Pannenberg, "Heilsgeschehen und Geschichte (Fortsetzung und Schluß)",

Here we see that Pannenberg's concept of *Rückbindung* incorporates the close interplay of two important concepts: the *continuing creation* of new and contingent events and God's *faithfulness* to those previous events already created, by connecting each new event not only back to the previous present but also (in a qualitative sense) with the physical tendencies and particularities evidenced by that past present. Over the course of time, in its movement from future to past, not only does God create each new moment as a free and contingent event, but due to his faithfulness to creation God also ensures that the physical qualities of each of these new moments exactly correspond with what he has previously brought into existence. For Pannenberg, it is this continual correspondence and coherence between contingent moments of God's future (a continuity stemming from God's faithfulness to his creation) which the natural sciences have then mistaken for an *inherent* continuity to the physical world. Whereas Newton spoke of the inherent nature of bodies to maintain their states (an active quality), Pannenberg understands such maintenance as direct evidence of God's faithfulness in the creation of each successive future moment. Thus we witness a process that is active only for God; the physical world exists only in a passive sense. Thus inertia "can be represented, with other forms of uniformity in natural occurrences, as an expression of God's faithfulness, the identity of which is based not on a natural unchangeability but on adhering to the decisions once made in each case".[132]

Pannenberg works here with a distinction between what are essentially active and passive roles. The classical view that physical matter progresses on its own from the past to the future under set laws of nature places the responsibility for temporal development on matter itself. But for Pannenberg, this shifted God's role in creation into a subordinate, passive role and opened the door to the dangers of deism and even the complete removal of God from the physical world. The principle of iner-

285, emphasis added.

132 Pannenberg, "Contingency and Natural Law", 109. That the book in the reader's hand has remained a book in the leap from the past minute into God's contingently-created present, and has not instead been transformed into a fish or has disappeared completely, testifies to the faithfulness of God to ensure that his newly-created present coheres exactly with the immediate past.

tia played "a decisive role for making nature independent toward its divine origin. The principle of inertia removed the philosophical demand for the constant cooperation of a first preserving cause for the preservation of the conditions of bodies and especially also of their movement".[133] Combined with this removal was the notion that such natural laws, as laws *per se,* worked in a general and universal manner leading to the strict uniformity of the physical world, without room for the contingent acts of a personal God. By using this shift to God's faithfulness in the process of a temporal *Rückbindung* of each contingent historical and temporal event, Pannenberg can now reclaim both the dominance of God in creation and also the personal activity of the divine which ensures the unity and uniformity of each of those events on a case by case basis. Thus nature becomes passively directed by the direct activity of God, and in the conflict between legalistic, universal "laws" of nature and the contingent and faithful decisions of a personal God, it is the personal creativity of the divine which wins out and provides time and nature with "the stamp of a personal power, not of a mere structure of laws".[134]

That the sciences believe they deal with universal laws at all is a presumption that can only arise on the basis of the faithfulness of God[135] who, despite the absolute freedom of his will, "bound himself through his contingent acting"[136] to the uniformity of his own acts.[137] To return to the example above: from Pannenberg's perspective, that the chemist expects magnesium oxide to be produced when burning magnesium in oxygen has nothing to do with any inherent qualities of the materials in their development from separated elemental magnesium and oxygen in their past states to a compound in their future state, since the movement of time *from future to past* rules out any such development from past to future. But that such an experiment has been repeatedly replicated and has "produced" the same results, would for Pannenberg in no way support the expectation of a stable and independent physical world. Rather

133 Ibid.
134 Ibid. 113.
135 Ibid. 109.
136 Ibid. 104.
137 "On the basis of the faithfulness of God, by his self-identification in the sequence of his contingent acting" (ibid. 109)—thus so-called "natural laws" and the continuity of events derive solely from acts of God.

these repeated experiments testify to the immense faithfulness of the creator God who refuses to allow future moments (despite their inherent and radical contingency) to be disconnected from the creation as God had once made it. To use a different metaphor, each moment sees God laying down the track of time; that each new piece of track should seamlessly connect to that piece of track which went immediately before it bears witness to God's faithfulness in sustaining his creation rather than derailing it.[138] Only in this way, in the combination of the absolute power and freedom of God in each new moment together with God's enduring faithfulness, can Pannenberg envisage a solution for theology to hold together the continuity of temporal events with an appreciation of their underlying and fundamental contingency.[139]

In this way, Pannenberg's approach finally overcomes all objections to God's direct action in history or in the processes of the natural world, while simultaneously accounting for the regularity of the experienced world. When God and time are envisioned this way, there can be no scientific or historical grounds to object to the historical or bodily resurrection of Jesus: that it goes against the principles of historical analogy or the lessons of experience (that dead people do not rise) merely points to the narrowness of our methodology and our expectations. That it contravenes "laws of nature"?—there are no such laws. Can we set limits upon God's power? Do we lecture God on what he may and may not do with

138 This would then be the great message of the flood narrative. Where once there was dry land to stand upon and air to breathe, there was suddenly only water. If each moment in time is a contingent act of God, then the land which sustains life one day can turn into the sea that destroys it the next. God is in no way constrained or obliged to ensure that each new and contingent moment of time *should* in fact connect with the last. God's creation, as *God's* creation, is thus subject only to God's continuing desire to sustain it. If God chooses no longer to sustain his creation, he brings into existence a new future moment and refuses to connect it to the last, derailing creation and beginning it again at a different point. It is God's covenant with Noah after the flood which highlights God's resolution never again to derail creation in such a way. This foreshadows Pannenberg's use of time as an instrument of God's judgement on both the physical and social worlds (discussed in the following chapter).

139 Pannenberg, "Heilsgeschehen und Geschichte (Fortsetzung und Schluß)", 285.

his creation and with his open future? Who are we that we should limit the freedom of God? What is history that it should limit the contingency of God's future and God's free will if he refuses to limit himself? And what is science that it should presume to draw boundaries for the *potentia absoluta* of God as *all-determining reality*?

Before I go on to offer some critique of this novel understanding of time, we need to examine one final aspect of Pannenberg's concept of time: the relationship between the present and eternity.

4.3.2 Controlling the Present

For Pannenberg, the way in which God relates to time is closely linked to God's power. We have already seen that Pannenberg understands God's power and sovereign rule as forming fundamental and indispensable aspects of God's being. Divine power is also at play in Pannenberg's understanding of the present. For him, the limits of the present are defined by the ability to express or maintain "control".[140] Rejecting a common scientific reduction of the present to a vanishing point between the future and the past, Pannenberg notes that "we do not at all experience the present only as a point" but rather "experience the present as a more or less extensive vicinity".[141] The key here is to understand what leads to variations in the size and "extension" of this now-shaped "vicinity". Maintaining his focus on the importance of power, Pannenberg's response is to link the extensiveness of the present to the individual's power for *control:* the "greater or lesser scope of the experienced present is connected with the specific possibility for acting and for taking a position. Everything over which we still exercise control and which we can still meet with decisions is counted as present".[142] The extension of "present" time for human consciousness is directly correlated with our ability to act. In practical terms, everything over which a person can still exercise control is counted as present, but once that person can no longer control an object or

140 This view appeared quite early in Pannenberg's career in the late 1950s and early 1960s during his time in Wuppertal and Mainz, and is reflected in his small anthropology from 1962. Cf. Pannenberg, *What is Man?*, ch. 6.

141 Ibid. 71.

142 Ibid. 72.

event then it is counted as past: "What we can no longer change lies for us in the past. That to which we cannot yet react is still future".[143] By defining the present tense in direct relation to the amount of control each individual wields, we see the point at which Pannenberg comes closest to developing a subjective understanding of time, that is, a non-unified or non-universal temporal system dependent upon the subjective consciousness of the individual. "The boundaries of the present are ... on the one side the unpredictable future and on the other side the past that is completed",[144] yet this past and future will vary from person to person and even within the experience of the same person in relation to different concurrent events.

While Pannenberg expresses this idea in active terms ("control"), in a "subjective" view of time we must take into account *all* aspects which may influence an individual's consciousness of time. This must include of course passively experienced events over which the individual lacks all control. The infliction of pain upon another (such as in the relationship between torturer and victim) provides a good example: the present for the torturer might be reduced to each individual cut or burn since the torturer acts in an active, "controlling" sense. The first burn quickly becomes a completed past event, as the second burn becomes the present. Yet for the victim, the "present" must be defined as that broader extensive field characterized by inflicted pain in a situation over which he or she precisely *lacks* any "control". For the victim, the present extends far beyond that period of time when the torturer acts; the victim remains in the present for as long as the wounds continue to burn, and for as long as the victim remains psychically caught up within that event (a period which may extend years after the event). In this respect, while Pannenberg speaks of the subjective present as defined by control (and uses active examples of such control: "boxing", "fencing"[145]), we must also allow the idea to extend over passively experienced states.[146]

143 Ibid.

144 Ibid.

145 Ibid.

146 As far as subjective temporal systems go, here we see the advantage of Müller's matrix of modalized time, in that it could account for the continued re-*present*-ation of past events into the subjective now. Cf. Müller, *Das unbekannte Land,* 218.

Interestingly, Pannenberg does not attempt to limit this subjective view of the present to *human* consciousness[147], but rather uses a double move which both extends it while simultaneously abolishing it. In his first move, Pannenberg projects this subjective understanding of the present onto God and God's own relation to time: thus what God no longer controls is counted as the past for God and likewise that upon which God must still wait is God's future. The twist here in Pannenberg's work is that such a situation remains purely unrealizable and hypothetical. The power of the all-determining God necessarily extends over all things.[148] Therefore, there simply can be no past or future for God, since all creation exists within the realm of God's power and direct control: everything stands "as an eternal present before the eyes of the person who has power over all things".[149] Having thus transferred this subjec-

147 Though it is questionable whether this aspect of control can be attributed at an anthropological level to the entire human race, or whether it would be more appropriate to claim this understanding of the present as an individual and cultural development. In literature studies and semantics, it is not at all clear that the shift from possessing the physiological capacity for seeing ordered events to then expressing these events via a narrative consciousness is an event which occurs at the genetic/anthropological level without first coming under the influence of culture. Cf. e.g. Micha Brumlik, "The Concept of Time and the Faculty of Judgement in the Ontogenesis of Historical Consciousness", in Jürgen Straub (ed.), *Narration, Identity and Historical Consciousness* (New York: Berghahn, 2005), 135–40. The difficulty is that once shifted into the realm of the personal, the validity of what counts as one's own "present", "past" or "future" becomes a matter for sociocultural approval or validation. "Depending on the sociocultural standards and rules, there can be 'obviously valid', arguable, and meaningless constructions of the past, present, and future, as well as arbitrarily complex temporal systems. ... The standards, criteria, forms, rules, and procedures a culture supplies its members for producing temporal systems of meaning are themselves products specific to a culture. A 'past', as a narrative construct shaped from a present perspective, is inextricably bound to the intellectual, linguistic, and communicative bases of its production"; Jürgen Straub, "Telling Stories, Making History: Toward a Narrative Psychology of the Historical Construction of Meaning", in id. (ed.), *Narration, Identity and Historical Consciousness* (New York: Berghahn, 2005), 44–98 at 80.

148 Pannenberg, *What is Man?*, 73 n. 4.

149 Ibid. 73.

tive view of the present onto the omnipotent divine, Pannenberg, in his second move, returns to the subjective, human perception of time in order to dismantle it. "The truth of time lies beyond the self-centeredness of our experience of time as past, present, and future".[150] In other words, having drawn on a subjective view of time in order to define the present, Pannenberg dismisses this approach by arguing that the true nature of time does not actually exist as a subjective structure of human consciousness. Rather it lies outside of consciousness as an objective reality. Indeed Pannenberg stresses that a subjective view of time based on the perceptions of our consciousness merely highlights the state of humanity's fallenness into the sin of "self-centeredness". The "perspective of our self-centered experience of time can consume our objectivity, which is open to the world. … As soon as that happens, the self-centeredness of our experience of time and space becomes a matter of being closed off. It becomes the source of the one-sidedness of our relation to the world and the sign of sin".[151]

Strictly speaking, this is not to argue that time is unrelated to a subjective frame of reference, but rather that the *true* nature of time is based within *God's own subjective frame of reference,* namely in God's eternal present, rather than in our own sinful experience of time.

4.3.3 Eternity as God's Frame of Reference

The concept of ether played such a vital role in the early stages of relativity physics because it offered the possibility of a universal frame of reference. Physics would not need to deal with the consequences of relativity if all physical measurements could be referred to the ether. In an analogous way, Pannenberg's shift of the "true" locus of time away from and out of subjective human consciousness and into the "objective" realm of the divine, in principle frees humanity from the absolute relativity and fragmentation of perceived time, since all time can now be measured or related to God's own universal frame of reference: God's eternal present. "The truth of time lies beyond the self-centeredness of our experience of time…. The truth of time is the concurrence of all events in [God's]

150 Ibid. 74.

151 Ibid. 77.

eternal present".[152] It is in this sense that Pannenberg can offer an understanding of time that has aspects of both objective and subjective models. For human beings, the true nature of time is objective ("the truth" of which lies outside human consciousness); but for God (or at least for an anthropomorphized projection upon God), time is subjective since it has its basis within God's eternal frame of reference, the eternal present constituted by God's omnipotent control.

Understandings of eternity in classical Greek philosophy and medieval Scholasticism placed eternity in opposition to temporality: temporality was the realm of change and potential, whereas eternity represented a realm lacking potential, unable to change or progress due to its perfection. "Ancient Greek philosophy understood eternity as that which always is and remains in the sense of what is universal".[153] A major problem then arose with the tendency to use the concept of eternity synecdochically to represent the being of God, transferring directly to the nature of God that which was thought regarding eternity. Again, this stems from usage in Greek philosophy, where constancy, completeness and simplicity were attributed to the nature of both the eternal and the divine.[154] This tendency was transferred to Scholastic theology via Aristotle, and made a strong appearance in Thomas Aquinas' discussions on the nature of God, specifically the issue regarding choice, resolve or action on the part of God. Scholasticism's difficulty with such concepts was that they imply movement or change for God and thus a temporality within the divine. Temporality was considered part of creation and thus properly external to the nature of God. If time did exist for God, then this would imply a degree of mutability and impermanence which, for Thomas, would lead to the awkward proposition that God is no longer perfect, since "any changing thing, whatsoever the change, is somehow potential", whereas "God [is] sheerly actual and unalloyed with potentiality". "Anything in change acquires something through its change, attaining something previously not attained",[155] whereas if God is truly perfect, then he needs nothing and will never change. Since God is understood to be self-sufficient and

152 Ibid.

153 Ibid.

154 Ibid. 75.

155 Thomas Aquinas, *Summa Theologiæ*, 1a, Q9, a1.

everlasting, theology witnessed the combination of divine nature with the concept of eternity.[156]

The difficulty for Thomas (as with other theists who continue to pursue this line) is an inability to present God in a satisfactorily trinitarian way. For example, in a proposed perichoretic trinity, where the three divine persons must act relationally, a refusal to act and react to others in open responsiveness destroys the life of the relation together with the persons involved.[157] Such relationality within the immanent trinity, the ability and desire to respond to the other, must presuppose a form of temporality within the nature of God, even if we can only speak of this temporality in a strictly metaphorical sense. While modern theologians may no longer find problems with such a metaphor of time within the divine life (based around relation, reaction and decision),[158] it was precisely this very concept of relationality, especially with regard to creation, that Thomas rejected or (at best) qualified as a "mixed relation" (that creation is related to God, but that God is not truly related to creation).[159]

Pannenberg correctly recognises that the opposition between eternity and temporality can lead to the danger of a dual cosmology, the creation of an ontic divide between divine and mundane realms which then in turn seeks to preserve God from efficacious contact with the realm of creation. Pannenberg opposes such an approach, arguing first that eternity "does not stand in contrast to time as something that is completely different. Eternity creates no other content than time. However, eternity

---

156 Langdon Gilkey, *Maker of Heaven and Earth: The Christian Doctrine of Creation in the Light of Modern Knowledge* (Lanham, MD: University Press of America, 1985), 56.

157 For this reason I cannot agree with the opposing position by Paul Helm, "Eternal Creation: The Doctrine of the Two Standpoints", in C. Gunton (ed.), *The Doctrine of Creation* (Edinburgh: T & T Clark, 1997), 29–46 at 46: "Immutability, immobility ... do not signal weakness and insufficiency in God, but the precise opposite, fullness of being. ... [God] does not need to change".

158 Cf. e.g. Moltmann's ability to speak of a form of divine temporality "before" the act of creation in "God's Determination of Himself to be Creator"; Moltmann, *God in Creation*, 79–86.

159 "Being related to creatures is not a reality in God, we say it about him because of the real relation in creatures"; Thomas Aquinas, *Summa Theologiæ*, 1a, Q13, a7.

is the truth of time, which remains hidden in the flux of time. Eternity is the unity of all time,[160] but as such it simultaneously is something that exceeds our experience of time".[161] Here again we find Pannenberg's implicit critique of dialectical and existential theology, specifically what he sees as their tendency to promote a dual cosmology, a division between temporality and the plane of God's eternity.[162]

Pannenberg then goes on to distance himself from the philosophical, theistic view of God which would prohibit God's activity and involvement in the temporal world,[163] arguing that the god of Greek philosophy has little to do with the Christian God: "the eternal God of the Bible is a living, active God",[164] one who can hold the entirety of time, including all its particularity, in his eternal present and rule over it with omnipotent control. By encapsulating all time and creation in the divinely controlled "eternal present", Pannenberg ensures that the Judeo-Christian God becomes "the God who works in history, in distinction to the god of Greek philosophy".[165] Thus in this sense, eternity and God's omnipotence become reflections of one another for Pannenberg, since only the God who operates within an unlimited frame of reference can be truly omnipotent: the "present that has a future outside itself is limited"[166] and can therefore not represent the realm of divine control and activity. It is

---

160 In his later work, Pannenberg uses the term "omnitemporality" to express this unification of all time into a single entity; cf. Pannenberg, "Eternity, Time, and Space", 102.

161 Pannenberg, *What is Man?*, 75.

162 See e.g.. Pannenberg, "Heilsgeschehen und Geschichte 1", 218, as well as his discussion on Barth's system of "pre-, super-, and post-temporality" in his *Systematic Theology*, i. 405–7.

163 The "god of Greek philosophy is not capable of doing anything. In order to act, God would have to pass from rest into movement and thus would lose his eternity along with his constancy. For the same reason, the god of Greek philosophy really had no attributes. Otherwise, along with his attributes there would be distinctions in God together with variability; thus again there would be no eternity. The eternity of the Greek god is that of the empty, eternal being"; Pannenberg, *What is Man?*, 75.

164 Ibid.

165 Ibid. 76.

166 Pannenberg, *Systematic Theology*, ii. 92.

for this reason that Pannenberg can group together the infinity of God, God's holiness, eternity, omnipotence and omnipresence into a single thematic complex.[167]

Pannenberg still operates with this idea in his later *Systematic Theology*, arguing that God's eternity is the true ground of all that which humans perceive as temporality.[168] Central here is perspective, that is the way that human beings and God differ in their perception and relation to time. "Eternity", as God's frame of reference, "is the undivided present of life in its totality".[169] This stands in contrast to "our human experience of time" which is limited in an active sense to the present only, and which must wait passively upon a future which is beyond itself. "We are not to think of [the eternal present] as a present separated from the past on the one hand or as the future on the other. Unlike our human experience of time it is a present that comprehends all time, that has no future outside itself. ... A present can be eternal only if it is not separate from the future and if nothing sinks for it into the past".[170] It is only the omnipotence of God which can maintain such a state. Thus Pannenberg's use of the divine, eternal present delimits itself from four other conventional views of time, each of which would then count as being overly focussed upon the "human perspective" on time. (1) God's eternal present is not the human "present", disconnected from the past and isolated from a future beyond itself; nor is God's eternal present (2) based upon a dualistic ontology which sees the eternal as distinct from and over against our own earthly temporal flux; nor (3) is God's frame of reference to be pushed only into the realm of the future and excluded there (separated from the present and past),[171] or (4) equated with process theology's developmen-

167 Cf. Pannenberg, *Systematic Theology*, i. 397–422.

168 Cf. ibid. 401–10, and Pannenberg, *Systematic Theology*, ii. 90–102.

169 Pannenberg, *Systematic Theology*, ii. 92.

170 Ibid.

171 Pannenberg rejects Georg Picht's critique of the concept of an "eternal present"; Georg Picht, "Die Zeit und die Modalitäten", in H. P. Dürr (ed.), *Quanten und Felder. Physikalische und philosophische Betrachtungen zum 70. Geburtstag Werner Heisenberg* (Brunswick: Vieweg, 1971), 67–76. In turn he critiques Picht's own focus "in favor of the primacy of the future in the understanding of time"; Pannenberg, *Systematic Theology*, ii. 92 n 238. Picht's view not only "isolates one of the three modes of time" but in so doing would (for Pannenberg)

tal process "from infinite to infinite".[172] Rather, God's eternal present is coterminous with the entirety of temporality, creating no other content than time itself, but merely revealing that which we *experience* as divided time as having its ontic basis in the unity of God's control, in the unity of God's eternal present. "We are not to think of eternity as the epitome of time. Rather, we are to think of time with its sequence of events—future, present, past[173]—as proceeding from eternity and constantly comprehended by it. Eternity is constitutive for the experience and concept of time. Only if time is basically a unity, i.e., as eternity, can we understand the nexus of that which is separated in its course".[174]

Eternity for Pannenberg can thus be understood along the lines of a first order category which provides the basis for human experiences and perceptions of time as separated and sequential second order categories. In the search for an external theoretical point of contact for this view, Pannenberg appeals to David Bohm's distinction between an implicate order and explicate order in temporal theory, where the implicate order acts as the origin of the evolving phenomena of the explicate order in natural phenomena,[175] reducing the expressed particularity of the explicate order to a secondary level. "The order of unfoldment at a given level emerges from a 'timeless' ground in which there is no separation".[176] Emergent sequence is thus preceded by a non-differentiated "timelessness": the ontic whole out of which the parts of our *perception* of time arise. While Pannenberg can speak here then (in the sense of objective time) of "the *prephysical*

isolate the activity and power of God.

172 Pannenberg argues that since process theology concentrates upon a temporal process based around atomistic events, it thus prioritizes the "part" over the "whole". Thus it reverses what Pannenberg sees as the philosophically and theologically necessary primacy of "whole" over "part"; cf. Pannenberg, "Atomism, Duration, Form", 113–129.

173 Note here that in his *Systematic Theology,* Pannenberg is still operating with the premises of *Rückbindung* developed in his earlier work, where "the sequence of events" in time flows from future to past.

174 Pannenberg, *Systematic Theology,* ii. 92–3.

175 Ibid. 92 n. 239.

176 David Bohm, "Time, the Implicate Order, and Pre-Space", in D. R. Griffen (ed.), *Physics and the Ultimate Significance of Time: Bohm, Prigogine, and Process Philosophy* (Albany, NY: SUNY Press, 1986), 177–208 at 196.

nature of time and its basis in eternity",[177] and of this prephysicality as the only pathway to solving paradoxes raised by the tendency of some physical theories "to derive the concept of time from specific physical data" (as with "Bohm's idea that we must view time as an abstraction from movement"),[178] we begin to notice the first hints of Pannenberg's own *quasi-divinization* of time as an attribute of divine power, and his slow removal of time from the proper realm of *creatures.* By elevating "the truth of time" to the divine frame of reference (away from fragmented human perceptions of temporality) time shrugs off its meaning precisely as something which is caught up within the fragmentation and brokenness of all creatures. To use a trinitarian metaphor, we are presented with an idealized "immanent" time, pure within God's own eternal frame of reference yet segregated and distanced from "economic" time with all its harsh, inexorable, deadly, decaying and fragmented particularity. Despite Pannenberg's desire to present eternity as containing "no other content than time", the particularity of time and its fallenness (as one of God's creations) becomes an illusion which is dramatically separated from the supposed truth of "time unified" in God's eternal present. Once purified of such destructive fragmentation and linked with the concept of omnipotent, divine control, this concept of unified time is ready not for redemption and fulfilment in God's new creation together with all other fallen creatures, but to be *instrumentalized,* to take on its role as a weapon in the hands of the absolutist and omnipotent God in the service of judgement on that fallen creation. Under such absolutist "rule by time", time serves as a mere tool for an omnipotent divinity who seeks to direct and control creation, yet also to suppress and punish it. The omnipotence of God as expressed via an eternal present defined in terms of absolute control, transforms time into the primary mechanism for achieving and maintaining God's practiced *potentia absoluta.*

4.4 Summary: Rule by Divine Decree

In Pannenberg's attempts to develop a new theory of time, we see evidence of his broad, theological programme to provide a theology with

177 Pannenberg, *Systematic Theology,* ii. 93 n. 239, emphasis added.

178 Ibid.

"universal validity".[179] Due to the dominant role played by relativity theory and the methodology and epistemology of the natural sciences, if Pannenberg's programme is to succeed, then he needs to display both the ways in which a *theological* concept of time overarches and includes the phenomena dealt with by the natural sciences, and the ways in which this new concept is able to overcome the difficulties and limitations of that scientific approach. In short, if Pannenberg hopes to replace the dominant theory of time with his own conception, he must show that his theological theory better accounts for the observed data while also addressing (or providing insights into) the problems left unanswered by the current physical theory. Pannenberg is certainly correct in his observation that "the underlying religious interpretation of reality is taken no longer as universally valid but as a matter of private preference, if not as superstition".[180] Yet what makes Pannenberg's attempt more difficult is his desire to ensure that his theory is at the same time thoroughly theological, and not just accessible to the natural sciences. In order to be considered "universally valid",[181] Pannenberg's concept of time must satisfy the theoretical claims of both science and theology. Fortunately, there is a significant amount of overlap at the philosophical level between physical theories of time and theological concerns. The practical insufficiencies of Minkowski space-time, for example issues of restriction, invariance and determination, are also concepts which have a long history of discussion within theology. Furthermore, Pannenberg's basic premise regarding the overlap of the natural world and God's creation already provides the overarching set within which the subset of physics and the physical world can be placed. While theology speaks of "creation" whereas the other sciences speak of "nature", this terminological difference should not "serve as a pretense to make acceptable the chasm between theology and knowledge of nature"; "If the one God of the Christian faith is also the Lord of nature, then we will not be satisfied with [such a] chasm".[182]

179 Pannenberg, *Anthropology in Theological Perspective,* 15.

180 Pannenberg, "Theological Questions to Scientists", 15.

181 In Pannenberg's limited sense of "universal", i.e. as satisfying the demands of the theological faculty and the physics department.

182 Pannenberg, "Contingency and Natural Law", 72. "If the God of the Bible is the creator of the universe, then it is not possible to understand fully or even appropriately the processes of nature without any reference to that God";

The challenge for Pannenberg is in the details: to find a way of providing a theological basis for the conceptions of an open future and for the existence of novelty; to provide a framework for change and development; to deal with the existence of non-substantial reality—and to do so (for the sake of his own intellectual integrity and further discourse with the natural sciences) without resorting to the postulate of an exclusive, inner theological "private sphere"[183]—and to achieve this all while still accounting for the experience of a fundamental regularity and predictability to the natural world. In short, Pannenberg's basic aim in his concept of time is *to find a way to unify radical contingency (as a reflection of the free and contingent will of God) with the phenomenological data of physical stability and predictable regularity.*

A range of factors were identified as impulses for Pannenberg's work towards this goal, beginning with his concerns regarding the dominance of causation and determinism in the underlying philosophy of the natural sciences, as well as the abandonment of contingency both in the natural sciences and in the philosophy of history (seen particularly in the analogical method). We saw how Pannenberg identified these factors as bearing a strong negative impact upon the "contents of the faith". Modernity views aspects of the faith with great suspicion. The leading example for Pannenberg is the resurrection of Christ, which is not seen as a radically novel and contingent act which breaks the bounds of the narrow epistemological preconceptions of the historical and natural sciences, but as a reflection of private superstition.

In this chapter we began by tracing Pannenberg's problem with the current state of affairs in the natural sciences, particularly with the concepts of inertia and natural laws. At a theoretical level, the special theory of relativity has come to dominate the discourse on time, and this dominance has proved so comprehensive that physical theories of time have become the *de facto* standard, even though in practice many still operate at the level of classical Newtonian physics, with its heavy focus on inertia and determinism.[184] Despite his desire to engage with the current

---

Pannenberg, "Theological Questions to Scientists", 16.

183 Cf. Pannenberg, *Anthropology in Theological Perspective*, 14.

184 Due to the high degree of mathematical and theoretical specialization required for working with relativity theory, it is not uncommon for those in the

science of time at the philosophical and theoretical level, we have seen that Pannenberg's major focus in his discourse with the natural sciences sadly remains limited to the Newtonian level: despite some reference to relativity, the force of his argument remains directed at the concept of inertia as the foundation of a deterministic natural lawfulness. As we saw, this weakens his contribution, and has attracted criticism from others in the discourse.

Yet Pannenberg was able to display how this deterministic sense of lawfulness within the natural sciences has gradually spread across academic disciplines into philosophy and history, and the repercussions which such a development has for theology. The great accomplishment of classical science was certainly its ability to replace the cult as the social institution able to tame the contingency, novelty, and uncertainty of the future. Whereas the cult presented a world open to the whims and contingent decisions of the gods, classical physics claimed not only to be able to determine what is naturally possible and what is not, but also to predict the future. In accordance with the lawful rules of development, a future event is merely the expression of that state of affairs which immediately preceded it. As the "flowering" of the past and present, there can be nothing radically new in the future, indeed nothing that was not already contained in the present and past. Once paired with a physical reductionism which excludes the existence of anything non-physical, then we are left with a strict determinism of both the natural world and the human person. This confluence of determinism, physical reductionism and the rejection of novelty formed the basis for Pannenberg's heavy critique of any system of development based on set, natural laws, regardless whether that development was posed within the natural sciences, within the historical sciences (as part of a historical morphology) or in philosophy (either in the concept of a Hegelian dialectic or process philosophy development). For Pannenberg, an "interpretation of the whole of history by means of the idea of development or any idea of an underlying teleology at all conflicts with the contingency of individual events and is therefore unacceptable on theological and historical grounds".[185]

---

practical sciences to operate still with a Newtonian model.

185 Pannenberg, "Heilsgeschehen und Geschichte (Fortsetzung und Schluß)", 262.

Pannenberg noted that the analogical method of the historical sciences was particularly insidious in its work to strip away all particularity from historical events, especially the particularity involved in the resurrection of Jesus.[186] This move is insidious because if the particularity of the resurrection is discounted from the start, then there is no way to present data which would challenge the current intellectual paradigm and lead to the required paradigm shift: "The possibility of the historicity of Jesus' resurrection has been opposed on the grounds that the resurrection of a dead person even in the sense of the resurrection to imperishable life would be an event that violates the laws of nature. Therefore, resurrection as a historical event is [understood to be] impossible".[187] Thus the concepts of inertia, determinism and development together with historical analogy form a matrix of natural lawfulness that prohibits belief in the historicity of the central events of the Christian faith.

Faced with this matrix, the easy solution would be to remove these faith events from the realm of history—an approach Pannenberg identifies and critiques in dialectic and existential theology. Yet Pannenberg's concern is that such a move would isolate the events of faith from our only means for objective evaluation, thus relegate them finally to the realm of superstition. Faith presupposes historical facts which are open to reason and rational historical enquiry and is not dependent merely upon the authoritarian statement of doctrine. "Whether or not a particular event happened two thousand years ago is not made certain by faith but only by historical research".[188]

Thus it was surprising to find that in the development of his own concept of time, Pannenberg falls back upon a position which can hardly be distinguished from a reliance upon "authoritarian statements of doctrine". In opposition to the matrix of inertia, determinism, development and analogy, Pannenberg introduces his own matrix built around *Rückbindung,* divine power, control, and an understanding of the eternal. The view that time flows from the future to the present, where it concretises, and then remains as the past, is an idea that if present throughout Pannenberg's work on time. The present is fundamentally determined by

186 Pannenberg, *Jesus—God and Man,* 89, 97.

187 Ibid. 98.

188 Pannenberg, *Jesus—God and Man,* 99.

the future; time runs from future to past and is maintained by divine direction and influence; the mutual exclusivity of causal, lawful natural development from past to future on the one hand, and the contingent freedom of the divine to introduce a contingent and novel future on the other[189], all remained constant themes in Pannenberg's work. Since Pannenberg was unwilling to surrender the historicity of divine actions or the events of faith, his response was to disassemble the scientific concept of deterministic natural laws and the concept of time upon which they are based. *Rückbindung* was an important tool for achieving that end, for arguing that *it cannot be the past which determines the present and the future, rather it is the future which must determine the present and the past.*[190] Reversing the flow of time ensures that there is no logical forward connection between past, present and future: the past is merely the remnant and residue of discrete, individual, contingent moments of the future successively placed into existence as our present. If the present is in fact determined by a future which is not yet existent (but resides solely with the contingent and omnipotent will of God) then there is no efficient way in which the present can be determined by past events. In this way, the bonds of inertia, development and determinism by natural law are broken, events can in no way develop or be predetermined by any pre-existing situation since each individual event has its origins in God's transcendent, free will.[191]

Five vital aspects are involved in the concept of *Rückbindung*. (1) *Sterile Linearity.* Since each atomistic moment of time is released from God's future and connected to the previous present by the will of God, nothing can grow from them, nor can they provide the basis for a "conception of a forward expanding development".[192] (2) *The Sovereignty and Rule of God.* The God who is able to achieve such a process of *Rückbindung* must be "the God who proves himself master over all",[193] the God "not condi-

189 Pannenberg, "Contingency and Natural Law", 83.

190 Pannenberg, "Theology and the Kingdom of God", 59.

191 Pannenberg, "Heilsgeschehen und Geschichte (Fortsetzung und Schluß)", 284–5.

192 Pannenberg, "Contingency and Natural Law", 106.

193 Pannenberg, "Theology and the Kingdom of God", 55.

tioned by anything",[194] the God who acts, but does not react,[195] the God who brings about each and every event in history,[196] the power that determines all reality[197]. (3) *Objectivity.* Despite human experiences of time, the "truth of time" is objective for human beings. It is external to human consciousness. The "truth of time lies beyond the self-centeredness of our experience of time as past, present, and future".[198] (4) *Contingency.* The radical contingency of all historical events (and God as the origin of that contingency) is the downfall of all "scientific *descriptions* of natural processes"[199] since "every event in which the future becomes finitely present must be understood as a contingent act of God".[200] (5) *Faithfulness.* Therefore it is only the faithfulness of God, and not the existence of deterministic or developmental laws of nature, or the natural persistence of objects, or inertia, which assures the existence of continuity in and between time, despite time's fundamental basis in contingency. When there is no forward-moving progression from past to future but only a rearward-moving flow of time from future to past, continuity in nature and history can only come from the faithful action of God.[201] In this way, Pannenberg side-steps that limiting matrix of inertia, determinism, development and historical analogy.

At least at an elementary, reductionistic, theoretical level (if not at a richer, externally coherent, theologically justifiable level), Pannenberg's

---

194 Cf. Pannenberg, "Contingency and Natural Law", 103–4.

195 Ibid.

196 "*Each event is something that happens to us* from the future which is until then still unrealized. It establishes anew the connection with that which happened earlier"; ibid. 85, emphasis added.

197 Pannenberg, "Contingency and Natural Law", 75. The "God of the Christian faith" must be "lord of the world ... the reality that determines everything"; ibid. 114 n. 5.

198 Pannenberg, *What is Man?*, 74.

199 Pannenberg, "Theology and the Kingdom of God", 67, emphasis added. Note that Pannenberg does not speak here of scientific *predictions* (which would be active and future-oriented), but of scientific *descriptions* (passive and oriented to the past).

200 Ibid. 61.

201 Pannenberg, "Heilsgeschehen und Geschichte (Fortsetzung und Schluß)", 285.

approach does overcome all objections to God's direct action in history or in the processes of the natural world. He does away with the scientific or historical grounds for objecting to the occurrence of contingent, divine acts in history. That events such as the resurrection go against the principles of historical analogy or the lessons of experience (that dead people do not rise) merely points to the narrowness of our methodology and our expectations; and it is futile to argue that such events contravene the "laws of nature", since Pannenberg has now done away with these "laws", replacing them with mere scientific "descriptions". For Pannenberg, if we set limits upon God's power, if via the natural or historical sciences we try to limit what God may and may not do with creation, then we risk abandoning a proper understanding of God, which Pannenberg sees inextricably and fundamentally linked with the concept of God as "all-determining reality".[202]

This concept of the omnipotent rule of God[203] feeds directly into Pannenberg's understanding of the present as an extensive vicinity of control.[204] While we saw that such an idea fails to do justice to the richness of human experiences of the present (e.g. in situations of pain and suffering over which the individual specifically has *no* control), Pannenberg extends this conception to God in order to argue three points. (1) What God no longer controls is counted as past for God. Likewise, that upon which God must still wait is the future. (2) Despite this definition, this situation remains completely unactualized for God, since the power of the omnipotent God extends over all things.[205] (3) Therefore all creation presents to God within an eternal present.[206] This omnitemporality[207] or eternal present ("the concurrence of all events") represents God's frame of reference (i.e. beyond the subjective perceptions of human temporality; beyond a limited human present) and is thus "the truth of time".[208]

202 Pannenberg, "Contingency and Natural Law", 75.

203 Pannenberg, "Theology and the Kingdom of God", 55: "God's being is his rule".

204 Pannenberg, *What is Man?*, 72.

205 Ibid. 73 n. 4.

206 Ibid. 73.

207 Pannenberg, "Eternity, Time, and Space", 102.

208 Pannenberg, *What is Man?*, 77.

The following chapter will examine in detail the problematic implications of Pannenberg's approach to time, including aspects merely hinted at here (such as the instrumentalization of time, and a reductionistic doctrine of God). However, in concluding this section, it will be helpful to foreshadow one point in the coming evaluation. Rather than overturning the problematic, deterministic matrix presented by the natural sciences, Pannenberg in fact embraces it. We have seen how the natural sciences used the tools of inertia and natural law, or the block universe and the four-dimensional manifold, to produce a stable and deterministic system bound under *rule by law:* the laws of inertial development, the laws of the conservation of energy, of the stability of chemical valencies, of nuclear, gravitational, thermodynamic or electromagnetic forces, etc. These all sought to define the progression of physical states in a deterministic way between one temporal moment and the next. In its minutest intricacies, the physical world was ruled *by* law: the laws of nature determined and specified exact progression through time. Rather than fundamentally overturning such a system based on rule by law, Pannenberg adopts the same system but replaces the concept of rule by deterministic, natural laws for rule by the deterministic will of an omnipotent God. Both systems remain linear; both systems remain objective; both systems remain deterministic. The essential difference is that Pannenberg's concept of time presents us with an absolutist *rule by divine decree* rather than a de-personalized *rule by natural laws.* Pannenberg's answer to the deterministic power represented by systems of natural laws is not an attack on the underlying concepts of determinism, but rather the attempt to outbid them, to trump them with the concept of an even more powerful determinant: the omnipotent God.

CHAPTER FIVE

THE PROBLEM OF HISTORICAL EVIL

5.1 Introduction

In his introduction to a collection of Pannenberg's essays "on Science and Faith",[1] Ted Peters hails Pannenberg as one of the most important voices in the modern science and theology discourse, noting that his "posing [of] theological questions to natural scientists" is both "startling and dramatic".[2] While Pannenberg's discussions with natural scientists are often dramatic, in this chapter we will see that Pannenberg's work on time extends upon a solid basis in medieval theology, and that his approach to the topic essentially reworks the thirteenth and fourteenth-century debate involving Averroistic determinism, the *via moderna,* and the reactionary *schola Augustiniana moderna.* Specifically, we will see that Pannenberg's work on time, with its stress on God's eternal present, control, and absolute omnipotent power, revives aspects of the medieval discussion of the *potentia Dei absoluta.* We will look at the particular relation between Pannenberg's approach and that put forward by Thomas Bradwardine, and suggest that Pannenberg's method suffers from the same failings seen in Bradwardine. Specifically, Pannenberg's concept of time under the *potentia Dei absoluta* leaves him with an inability to deal well with theodicy and the phenomena of sin and suffering, a failure to maintain a satisfactory *theologia crucis* before an overweighted (and politically absolutist) *theologia gloriae,* and a difficulty escaping the tension

1 Wolfhart Pannenberg, *Toward a Theology of Nature: Essays on Science and Faith* (Louisville, KT: Westminster/John Knox Press, 1993).

2 Ted Peters, "Pannenberg on Theology and Natural Science", in Wolfhart Pannenberg, *Toward a Theology of Nature: Essays on Science and Faith,* ed. Ted Peters (Louisville, KT: Westminster/John Knox Press, 1993), 1–14 at 1.

between determinism and human free will. In addition to these considerable difficulties (which Pannenberg shares with Bradwardine and with Luther, who follows Bradwardine in many respects), this chapter will also suggest that Pannenberg finally instrumentalizes time as a tool of God's judgement upon creation, rather than itself being seen as a creature in need of redemption and fulfilment.

5.2 Excursus: Averroistic Determinism and the Defence of Contingency

On the surface, Pannenberg's discussion with physics on the nature of time seems like a thoroughly modern topic, yet at its heart it revolves around a debate on determinism and contingency that is centuries old. Thus in responding to this "modern" dilemma, Pannenberg reclaims significant theological traditions from the late Middle Ages. Indeed, it is often forgotten that Pannenberg began his professional theological career as a medieval scholar and was well known in his early years for his competence in medieval and scholastic theology.[3] Hence, it should not

3 Cf. e.g. Pannenberg's doctoral thesis, *Die Prädestinationslehre des Duns Skotus im Zusammenhang der scholastischen Lehrentwicklung* (Forschungen zur Kirchen- und Dogmengeschichte, 4; Göttingen, 1954); Wolfhart Pannenberg, "H. Mühlen: Sein und Person nach Johannes Duns Scotus, Werl 1954", in *Theologische Literaturzeitung,* 80 (1955): 353–5; Wolfhart Pannenberg, "Johannes Duns Scotus", in H. Brunotte and O. Weber (eds.), *Evangelisches Kirchenlexikon* (Göttingen: Vandenhoeck & Ruprecht, 1956), 980–2; Wolfhart Pannenberg, "B. Hägglund: Theologie und Philosophie bei Luther und in der ockhamistischen Tradition, Lund 1955", in *Archiv für Reformationsgeschichte,* 47 (1956): 273–5; Wolfhart Pannenberg, "Albertus Magnus: Opera Omnia Vol. XXVIII (ed. Bernhard Geyer), Münster 1951", in *Theologische Literaturzeitung,* 81 (1956): 109–10; Wolfhart Pannenberg, "J. Duns Scotus: Opera Omnia Vol. I–III (ed. Carolus Balic), Civitas Vaticana 1950/54", in *Theologische Literaturzeitung,* 81 (1956): 550–2; Wolfhart Pannenberg, "Th. Steinbüchel: Vom Menschenbild des christlichen Mittelalters, Basel 1951", in *Theologische Literaturzeitung,* 81 (1956): 733–5; Wolfhart Pannenberg, "Anselm von Canterbury: Cur Deus Homo. Warum Gott Mensch geworden. Besorgt und übersetzt von F. S. Schmitt, München 1956", in *Theologische Literaturzeitung,* 82 (1957): 203–4; Wolfhart Pannenberg, "Akt und Sein im Mittelalter", in *Kerygma und Dogma,* 7 (1961): 197–220; among many others.

be surprising to find Pannenberg appealing back to the insights of the medieval church in order to address what some have mistaken as a peculiarly recent problem.

5.2.1 Natural Determinism and the Averroisitc Threat

The late establishment of Christianity's place in the Roman Empire (begun by Constantine and completed under Theodosius I) set the stage for an unopposed superiority of Christian thought and doctrine throughout the Roman world, a superiority finally actualized in 529 under Justinian. "Eusebius's lofty view of the imperial office became full reality in the person of Justinian…: one empire, one church, both led by one emperor",[4] using one prescribed intellectual system, Christian thought. A direct expression of this imperial attitude came in Justinian's 529 closure of the Platonist Academy in Athens—an event "filled with religious and intellectual meaning".[5] "The symbolism of the gesture was pregnant. It meant either that the ancient learning was superseded by Christianity, or that it was absorbed by it. In the disappearance of the Academy the last elements of paganism in the form of Hellenism were ousted … Paganism then had to be silent or go underground or branch out … [seeking] refuge in northern Mesopotamia before, it is thought, meeting in the eighth century the Arab scholars of Baghdad, who thus became the guardians of Greek philosophy through the Middle Ages".[6]

Under the Byzantinism of Justinian, imperial power was unified with the politically-supported intellectual dominance of Christian doctrine. Given this unification, it is not surprising that the decline and fall of the empire to Arab control, and the subsequent influence of (Hellenic-flavoured) Arabic intellectualism, would represent a major threat to a Christian academy striving to maintain an intellectual tradition while surrounded to the west, south and east by stronger, imperial Islamic forces.[7]

---

4 Gérard Vallée, *The Shaping of Christianity: The History and Literature of its Formative Centuries (100–800)* (New York: Paulist Press, 1999), 118.

5 Ibid.

6 Ibid. 118–119.

7 Under the Umayyad caliphs (660–750) and the subsequent Abbasid

By the time of the late-thirteenth century, the "intellectual purity" envisioned by Justinian had been replaced through the rediscovery and reception among the scholastics (notably Albertus Magnus and Thomas Aquinas) of Aristotle's philosophy, via Arab commentators such as Averroes and Avicenna. However, it would be the broader consequences of Averroes' work as a naturalistic, quasi-scientific method which would represent a significant concern for theologians at the end of the thirteenth century.

The major problem presented by Averroism was "a kind of determinism in which God need not be directly involved in the world".[8] "Islamic philosophers made God too impersonal, too far removed from his own creation"[9] since that creation (including the created human will) followed set, deterministic natural laws, removing the need for any direct and unmediated divine involvement in the course of the world.[10] Thus Christian thought was faced with a powerful philosophical theory with strong academic representation in Paris by figures such as John of Jandun, which essentially confronted the Christian belief in the unmediated action of God in nature and history. As Pannenberg notes, both objects

---

caliphate (750–1258), "Christendom" was reduced to a small enclave. After "the *pax Islamica* was established ... entire regions previously divided between Rome and Persia were now joined together as a result of the conquest. Suddenly, the world woke up to realize that a momentous event had taken place: 'The largest empire ever created in ancient times' had just emerged"; ibid., 201. P. Brown, *The Rise of Western Christendom: Triumph and Diversity* AD *200–1000* (Cambridge, MA: Blackwell, 1996), 8; "Made up of a galaxy of ancient lands, that stretched from Andalousia to Turkestan, the Islamic empire of Caliph Harun al-Rashid (788–809) dwarfed the empire of his contemporary Charlemagne ... recently established at the northwestern extremity of Eurasia".

8 Edith Wilks Dolnikowski, *Thomas Bradwardine: A View of Time and a Vision of Eternity in Fourteenth-Century Thought* (Leiden: Brill, 1995), 190.

9 Ibid.

10 Pannenberg, "Bradwardine and the Pelagians", 355–61. Both the Ockhamists and Bradwardine "verstehen das Verhältnis von Gott und Geschöpf ... als akthaftes und unmittelbares Willensverhältnis. Damit stehen beide nicht nur zur metaphysischen Theologie des 13. Jh., sondern vor allem zum arabischen Determinismus der Averroisten, zu Schicksalsglauben und Astrologie in Gegensatz'; ibid., 357.

and persons alike were subject not to the will of God but to a cosmically mediated determinism via necessary cosmic causes and through "the compulsion of finite objects" [*Zwang durch endliche Dinge*] that surround them.[11]

One loose group of theologians attempted to deal with the strong natural determinism of Averroism. The so-called *via moderna* or nominalist school—with representatives in Paris, Heidelberg, Erfurt and Oxford (including Ockham and his students Petrus Aureoli, Adam of Wodeham and Robert Holcot, as well as Pierre d'Ailly and Gabriel Biel)[12]—sought to counter deterministic Averroism by stressing the importance of human freedom, and thus moral responsibility. Yet the tendency of the group, and particularly Ockham, to argue that this freedom was displayed in the human ability to initiate their own justification by "doing their best" (*facienti quod in se est Deus non denegat gratiam*) quickly saw members of the movement branded as "modern Pelagians".[13] The threat of Pelagianism provided impetus for a backlash and counter-movement from what would later become the *schola Augustiniana moderna* (spurred on in England by Thomas Bradwardine and John Wycliffe,[14] as well as by Gregory of Rimini in Paris).[15] Despite their opposition to the *via moderna* on the issue of justification, the two groups were nonetheless in agreement regarding the need to defend against the deterministic theories of Averroism. Yet members of the *schola Augustiniana moderna* sought to

11 Ibid.

12 Ibid., 358; for an introductory view, cf. Alister McGrath, *Christian Theology: An Introduction* (Oxford: Blackwell, 2001), 38.

13 A 1326 commission of six theologians in Avignon censured fifty-one articles from Ockham's writing, though McGrath suggests the commission's critique represented "a culpable misunderstanding of Ockham's intentions"; cf. Alister E. McGrath, *Iustitia Dei: A History of the Christian Doctrine of Justification* (3rd edn.; Cambridge: CUP, 2005), 153–4.

14 For context (including the period directly following), see the introduction and selection of primary sources in David Charles Douglas, *English Historical Documents* (Abingdon: Routledge, 2004) 589–638, esp. 627–8; and 880–7.

15 Gregory of Rimini is generally identified as the founder of the *schola Augustiniana moderna*. But since Bradwardine predates Gregory, it can be argued that the origins of the school actually lie in Oxford rather than Paris; cf. McGrath, *Iustitia Dei*, 173.

achieve this end without resorting to what they considered overtly Pelagian arguments.[16] As a result they critiqued Ockham and the nominalists by arguing that all soteriological resources were located exclusively outside human nature (taking issue with the formula *facere quod in se est*): "Even the ability to desist from sin and turn to righteousness arose through the action of God, not through human action".[17] Here Bradwardine would play an important role.

5.2.2 Bradwardine's Solution

If a focus on human freedom is to be rejected out of fear of Pelagianism, how can a defence still be raised against Averroistic natural determinism? The first hint comes from the *via moderna* itself, as expressed in the work of Ockham's student Adam of Wodeham.

Adam of Wodeham extended Ockham's sceptical epistemology, arguing against natural causation and determinism by supporting the radical contingency of future events (including divine revelation) based on the freedom of the *divine* rather than human will.[18] He saw God's will as so radically absolute that contingency ruled even within the divine person: God was free to reverse and invalidate his own revelations and, if he chose, even lie and sin.[19] Furthermore, God's contingent will was

16 Achieving this, ironically, by pushing a view of double predestination that went further than Augustine himself was prepared to go; cf. McGrath, *Iustitia Dei*, 171–72. "Bradwardine supports his anti-Pelagian theology with a metaphysical doctrine of divine omnipotence quite alien to Augustine ... It is for this reason that some have questioned whether Bradwardine may be viewed as 'Augustinian' in the strict sense of the term. ... The fundamental difference between Gregory and Bradwardine is that the former managed to develop a theological stance similar to that embodied in Augustine's anti-Pelagian corpus within the the context of the intellectual framework of the *via moderna* where Bradwardine followed the *via antiqua*"; Alister McGrath, *The Intellectual Origins of the European Reformation* (Oxford: Blackwell, 2004), 86–7.

17 McGrath, *Christian Theology*, 39.

18 Pannenberg, "Bradwardine and the Pelagians", 358: "Adam of Woodham [gründete] die Kontingenz des Zukünftigen auf die Freiheit nicht des geschöpflichen, sondern des göttlichen Willens".

19 Gordon Leff, *Bradwardine and the Pelagians: A Study of his "De Causa Dei"*

so absolute that it was neither bound nor corresponded to natural laws: "a contingent event thus understood is characterized by the fact that *it stands in no necessary connection with the previous event* (at least in so far as it concerns God's *potentia absoluta*)".[20] Importantly foreshadowed here is the stress on the particularity of contingent historical events brought about by the free will of God, events which have no necessary or determinant connections between them.

While such an appeal to God's *potentia absoluta* as the basis of contingency was not specific to William of Ockham's school, it was extremely valuable in the defence of created contingency and divine freedom in light of the threats posed by the natural determinism of Averroism,[21] and it is here that we begin to see its relevance for the modern context Pannenberg is addressing. Taking his lead from Étienne Gilson[22] and Paul Vignaux[23], Pannenberg notes that:

> Interest in the *potentia absoluta* had already grown during the last quarter of the thirteenth century, long before Ockham's founding of nominalist epistemology. The reflections on God's absolute power served *to defend the sovereignty of God* from Averroistic determinism, which dissolved God's work within the mechanistic connections of the world order, and rendered God himself a mere exponent of this world order. ... Reflections on the *potentia absoluta,* the radical contingency of all created things—not only the entire world but also *each individual event in itself—the independence in the world process of each part from all other parts,* thus also established the contingency of each part within the order

and its Opponents (Cambridge: CUP, 1957), 250–4. Pannenberg, "Bradwardine and the Pelagians", 358: "so mußte Adam seinerseits die Konsequenz ziehen, daß Gott und Christus uns mit ihren Voraussagen trotz bester Absicht faktisch täuschen könnten, da Gott sich in der Zwischenzeit bis zum angekündigten Eintreffen der Voraussage gleichsam noch anders entschließen kann".

20 Pannenberg, "Bradwardine and the Pelagians", 358, emphasis added.

21 As with others, William of Ockham and his students wanted "in their own way to bring the sovereignty of God to bear over against Averroism"; Pannenberg, "Archbishop Thomas Bradwardine", 174.

22 É. Gilson, *Jean Duns Scot. Introduction a ses positions fondamentales* (Paris: Vrin, 1952), 77ff., 272ff.

23 P. Vignaux, *Nominalisme au XIVe siècle* (Paris: Vrin, 1948), 22ff.

> of salvation. ... [O]*nly through God's free discretion ... are the individual parts each endorsed and then slotted together into an order.*[24]

At this point, we see on the one hand a powerful concept of natural, cosmic determination through finite objects, and on the other a matrix of doctrinal points which seek to overcome the power of such a deterministic system: the radical contingency of all events, the particularity of historical phenomena, their freedom from natural determinants due to their existence through the free will of God, the *potentia Dei absoluta* which acts as the foundation of the radical contingency of events and the contingent absolute will of God who binds these individual, contingent events into a coherent order. Yet it would be the important contribution of Thomas Bradwardine to collect this diffuse matrix of doctrinal points together into a unified system and to push the logic of this approach to its ultimate end.

In his 1344 *De causa Dei,* Bradwardine presents a polemic against attempts to limit the absolute power of God.[25] Thus the *potentia Dei absoluta* and the related stress on the sovereignty of God represented for Bradwardine not only a doctrinal starting point but also the goal of his defence. It also supported the criterion by which his opponents could be judged: "Bradwardine's most obvious opponents were those ... Islamic philosophers who promoted a kind of determinism in which God need not be directly involved in the world".[26] Yet Bradwardine also rejected those who "went too far in the opposite direction ... by emphasizing the large scope for free will in human choice".[27] According to Pannenberg, Bradwardine instead pushed the concept of the *potentia Dei absoluta* to its logical end: to render contingent *all natural events* as well as *all human events* under the supreme sovereignty and power of God, since it was alone by the power and will of God that the contingent event could exist. Bradwardine's concern was to challenge dependence upon "autonomous natural experiences", instead perceiving the human being completely

24 Pannenberg, "Bradwardine and the Pelagians", 359, emphasis added.

25 Dolnikowski, *Thomas Bradwardine,* 165. Dolnikowski notes that despite Bradwardine's conflict with Ockham and his vehemence against so-called Pelagian concepts, the main thrust of his polemic was directed at any theological position which would limit the absolute power of God.

26 Ibid. 190.

27 Ibid.

from the perspective of the omnipotent God.[28] Dolnikowski notes that Bradwardine did not really reject the concept of determinism *per se* (his own conception of God being heavily dependent upon Greco-Arabic conceptions),[29] but rather his concern was with the directive force behind such a determinism. While his work shares with Duns Scotus and Ockham the desire to stress the unmediated nature of divine activity in creation over against Arab determinism,[30] and the power of divine activity in human affairs via the doctrine of double predestination, Bradwardine radicalises the approach by swapping out "determinism by natural forces" for "determinism by God":

> he simply replaces a cosmically mediated determination, operating through the powers of the cosmos, with an unmediated determination through God himself. Even the human being, while free from compulsion through finite things, is in each of life's movements completely subjugated to God [*Gott gänzlich unterworfen*]. For Bradwardine, there remains no room for creaturely independence. ... God [is] the decisive, driving force in all creaturely events.[31]

If the *potentia Dei absoluta* is to be taken seriously as a ground for the ultimate sovereignty of God over creation, then there can be no fanciful division between created *natural* objects and created *human* objects: as creatures, they are all subject to the will, power and sovereignty of God. Nor is the power and sovereignty of God to be seen as temporally limited. Bradwardine uses the concept of God's "eternal present"[32] to stress the lack of any objective difference for God between past, present and future.

> God's will simultaneously directs all things which we see modally as past, present and future. Therefore, God is not bound by our perception that a future event finally occurs in time, passes into the past, and can no longer be changed. Because God is free from the temporal limits of succession, he does not have to regard past and future as different. The truth of a past event, like the truth of a future event, is confirmed by God's

28 Cf. Pannenberg, "Bradwardine and the Pelagians", 357.
29 Dolnikowski, *Thomas Bradwardine,* 190.
30 Pannenberg, "Bradwardine and the Pelagians", 357.
31 Ibid.
32 Cf. Dolnikowski, *Thomas Bradwardine,* 184–91.

> will alone.... Though we see the future as contingent and the past as unchangeable, and therefore necessary, God assures the truth and reality of both in his eternal present.[33]

In this way, Bradwardine provides a unified theory which simultaneously rejects both Averroistic determinism and the perceived dangers of Pelagianism. Subjected to God's absolute power, all creaturely events, including human events, represent radically contingent and independent historical phenomena, dependent solely upon the free and illimitable will of God. As radically contingent instances, their freedom from natural determination is assured by both the will and power of God, and it is the *potentia Dei absoluta* and the contingent absolute will of God which binds these individual, contingent events into a coherent order.

It may at first seem counter-intuitive that Bradwardine uses divine determination to defend the existence of contingency, but we need to be clear on what Bradwardine is offering: it is not the rejection of all determination, but rather a change of lordship. The omnipotent God is the only Lord of the cosmos and thus it is God's will which is the source of all creaturely, temporal events. Thus both the human person and the greater creation are freed from the mechanistic processes of the world-order. Equally, it is only the absolute power of God over each and every contingent instance which defends God's true sovereignty and allows for his unmediated activity within creation. Metaphorically, in the shift to a radical, divine determinism human beings go from being slaves to the world to being slaves to God.

Bradwardine's work took the concentration on God's absolute power to its logical conclusion, leading to two significant and unavoidable problems. The first is seen in the criticism raised by Pannenberg, that this tendency of God's absolute power over past, present and future human events removes all independent self-determination from creatures, leading to a theology of strict divine determinism. Dolnikowski attempts a defence of Bradwardine on this point, arguing that Bradwardine's concept of God's eternal present need not necessarily lead to unavoidable determinism, which was an end Bradwardine sought to avoid.[34] But once

33 Ibid. 184.

34 Ibid. 191–2.

the concept of God's eternal present is combined with God's absolute power and will, and then further solidified through an appeal to the contingency of all events under the *potentia Dei absoluta,* the conclusion becomes a hard one to avoid. This is certainly the line that the young Pannenberg followed when he argued that it is not enough to maintain that Bradwardine was conscious of the threat of determinism in his own system, if he was finally unable to avoid it. That "Bradwardine did not advocate a theological determinism ... [is a thesis] one can hardly follow. While it is clear that [Bradwardine] attempted to avoid the deterministic consequences of his doctrine of God", nevertheless "the force of his systematic approach thrust him to deterministic consequences against his own will".[35] Curiously, one of the significant concerns that Averroism raised for Christian doctrine was its use of the principle of determinism to avoid personal responsibility.[36] Yet by stressing the *potentia Dei absoluta* and using it to develop a doctrine of double predestination, Bradwardine himself leads theology to this end. His attempts to preserve human freedom against the heavy determinism of his own system (such as with his doctrine of the *libertas contradictionis*[37]) buckled under the weight of the *potentia Dei absoluta.* Pannenberg perceives here only the semblance of freedom rather than its reality: human beings can act *as if* they were free, only due to their ignorance about which decision God has ultimately made for them.[38]

---

35 Pannenberg, "Archbishop Thomas Bradwardine", 175; also: "die logischen Konsequenzen drängen hier zum Determinismus" (ibid.). Of importance for our discussion of Pannenberg's concept of time is Pannenberg's own understanding of Bradwardine's system and its limitations; therefore whether Dolnikowski's view of Bradwardine is correct is an issue that we need not evaluate here.

36 Cf. Ian Maclean, "Heterodoxy in Natural Philosophy and Medicine: Pietro Pomponazzi, Guglielmo Gratarolo, Girolamo Cardano", in John Brooke, Ian Maclean (eds.), *Heterodoxy in Early Modern Science and Religion* (Oxford: OUP, 2005), 1–30 at 9.

37 H. A. Oberman, *Archbishop Thomas Bradwardine. A Fourteenth Century Augustinian. A Study of his Theology in its Historical Context* (Utrecht: Kemink en Zoon, 1957), 67, 113.

38 Bradwardine's attempts to hold to "a psychological *libertas contradictionis* for the human being ... [consists] strictly speaking only in the person's

This divine determinism leads directly to the second, more significant problem with Bradwardine's system. If there is no option for human freedom, if the acts of human beings are directly determined, stemming from the contingent will of God (as expressed through God's absolute power and predetermination), then it is difficult to avoid the conclusion that God is the final responsible cause for evil. A strict focus upon the *potentia absoluta* in Bradwardine's theology teamed with the radical contingency of all events makes it awkward to conceive of sin and evil as related to human action. Significantly, Bradwardine himself was deeply aware of this dilemma and searched for a means to avoid it. He tried to locate the origin of evil in secondary causes, "so that God may be said to predestine *to* evil, but not to *predestine* evil",[39] but even this attempt could hardly evade the ultimate conclusion. "The principle pushes us even to trace sin back to God as its author, and to understand God's will as the ultimate cause even of damnation".[40] Whether Bradwardine was content with the conclusions to which his system chained him does not detract from the conclusions themselves. Pannenberg perceptively notes that "although this trend stands in tension with [Bradwardine's] intentions, as his various attempts to stave off the logical consequences of his approach reveal", his theological motives "do nothing to change the fact that the systematic-logical structure of his approach has a deterministic character", one that creates serious problems in his doctrine of God regarding the "origin of evil".[41]

5.3 Resurrecting the *potentia Dei absoluta*

The parallels between Bradwardine's use of the *potentia Dei absoluta* and Pannenberg's own appeal to the absolute power, rule and sovereignty of God are clear. Even the parallels between Bradwardine's context and Pannenberg's understanding of the theological discourse with the natural sciences are strikingly similar. With Averroism, Christian theologians

ignorance about that which God has decided about him ... : he acts *as if* he were free"; Pannenberg, "Archbishop Thomas Bradwardine", 175.

39 McGrath, *Iustitia Dei,* 172.

40 Pannenberg, "Bradwardine and the Pelagians", 357.

41 Pannenberg, "Archbishop Thomas Bradwardine", 175.

of the thirteenth and fourteenth century faced an influential, academic system arguing for a natural determinism[42] which threatened the sovereignty of God,[43] effectively limited (if not excluded) the immediate activity of God in the natural world,[44] and allowed for a naturalistic course of development distanced from divine direction,[45] while at the same time the realism of the early scholastic period focussed on the discovery of universal and eternal truths to the detriment of historical particularity.[46] In response, the nominalists attempted to shift the focus away from historical "ideals" to emphasise historical particulars,[47] and appealed to the *potentia Dei absoluta* as the basis for all such contingent, historical particulars[48] and their coherence in the natural world.[49] Thomas Bradwardine then brought this dependence upon the *potentia Dei absoluta* to its fulfilment by extending its jurisdiction (as truly "absolute") to the human being,[50] and its reach to all temporality (past, present and future) by appeal to God's "eternal present",[51] and in so doing defended both God's sovereignty and power as well as the historical particularity and contingency of events.

Six centuries later, Pannenberg sees modern theologians facing the dominant scientific systems of Newtonian and relativity physics which argue for a natural form of lawful determinism either via the principle of inertia (in classical physics)[52] or the invariable space-time manifold of relativity theory[53]. In the former, "the first classic conception of natu-

---

42 Dolnikowski, *Thomas Bradwardine,* 190.
43 Pannenberg, "Bradwardine and the Pelagians", 359.
44 Ibid. 357; Dolnikowski, *Thomas Bradwardine,* 165.
45 Pannenberg, "Bradwardine and the Pelagians", 357.
46 For an overview of the debate between realism and nominalism in the context of the broader discussion on the *potentia Dei absoluta,* see William Placher, *A History of Christian Theology* (Philadelphia: Westminster, 1983), 162–80, esp. 162–7.
47 Ibid. 163.
48 Pannenberg, "Archbishop Thomas Bradwardine", 174.
49 Pannenberg, "Bradwardine and the Pelagians", 359.
50 Ibid. 357.
51 Dolnikowski, *Thomas Bradwardine,* 184.
52 Pannenberg "Contingency and Natural Law", 109–13.
53 Stannard, "God in and beyond Space and Time", 114.

ral science that emerged in the modern era ... man [feels] himself to be living in a world of eternal and unbreakable laws which [govern] his own life as well as the realm of nature",[54] which also effectively limit if not completely exclude the activity of God in the natural world, instead providing the basis for a completely naturalistic form of historical and cosmic development lacking any contingency.[55] The idealistic focus on universal truths in the form of natural laws led to a scientific rejection of the historically particular, excluding the data of particular phenomena when they fail to be assimilated into the predictive systems of natural laws.[56] This epistemological idealism spread to the historical sciences, which then adopted the analogical method, ruling out all instances of historical novelty or particularity, including the resurrection of Christ, when they fail to correlate with similar historical events.[57]

In Pannenberg's interpretation and analysis of the current discourse between the natural sciences and theology, there is a fundamental underlying similarity and parallel between these two historical contexts. It is therefore not surprising that Pannenberg returns to the thought world of these medieval scholastics, reclaiming the central, doctrinal importance of the *potentia Dei absoluta* and instituting a *neo-Bradwardianism.* In Pannenberg's use of the concept of *Rückbindung*—the contingent flow of events from future to present to past[58] placed in coherent order by the all-determining and omnipotent God[59]—we have the revival of a stress on the *potentia Dei absoluta* and the insufficiency of natural order and natural determinism: the inability of history or nature to provide coherent development in time;[60] God's power to bring into existence (from his future) and to maintain otherwise naturally disconnected, radically contingent and historically particular events;[61] God's power to link these events into a coherent order in accordance with his power and

54 Pannenberg, "The Biblical Understanding of Reality", 17.

55 Eddington, *The Nature of the Physical World,* 286.

56 Ibid.

57 Pannenberg, "Heilsgeschehen und Geschichte (Fortsetzung und Schluß)", 262–7.

58 Pannenberg, *Systematic Theology,* ii. 92.

59 Pannenberg, "Heilsgeschehen und Geschichte (Fortsetzung und Schluß)", 285.

60 Pannenberg, "Contingency and Natural Law", 83.

61 Pannenberg, "Theology and the Kingdom of God", 59.

faithfulness;[62] the absolute power of a God who expresses absolute control over all time from the realm of his omnipotent, eternal present;[63] and the absolute sovereignty of a God whose power defines his very being[64]—whose *"being is his rule"*.[65] In a mirror of Bradwardine's attack on a powerful, deterministic Averroism and a nominalistic "Pelagian" focus on the freedom of human beings, Pannenberg resurrects the spectre of the *potentia Dei absoluta* and releases it against the natural sciences in order to defend the idea of a God who intimately and immediately controls the entirety of nature and directs all the events of human history.[66]

Yet given Pannenberg's own expertise in late medieval theology and his intimate familiarity with Bradwardine's work, it is surprising that he adopts Bradwardine's approach without considering the significant drawbacks and dangers. The young Pannenberg had argued that Bradwardine's overwhelming focus on the *potentia Dei absoluta* in both nature and in human history led to an unavoidable divine determinism and the final attribution of sin and evil to God. Seven centuries later, can Pannenberg's work avoid these results?

5.3.1 Theodicy as Insurmountable Hurdle

Two core problems arose from Bradwardine's use of the *potentia Dei absoluta:* (1) the ultimate attribution of evil and sin (and thus suffering) to God, and (2) the discounting of human free will. I would argue that these two problems will be evident in any approach that seeks to follow Bradwardine and build on God's overwhelming *potentia absoluta*. Indeed, as we will see below, Luther also draws on Bradwardine's theology and also displays the same difficulties. However, as the concept of human

62 Pannenberg, "Heilsgeschehen und Geschichte (Fortsetzung und Schluß)", 285.
63 Pannenberg, *Systematic Theology*, ii. 92.
64 Pannenberg, "Contingency and Natural Law", 114 n. 5.
65 Pannenberg, "Theology and the Kingdom of God", 55, emphasis added.
66 "Every event in which the future becomes finitely present must be understood as a contingent act of God who places that finite reality into being"; ibid. 60. "If theology wants to think of the deity of God, then it has to think of God as the power that determines ... human history [and] nature"; Pannenberg, "Contingency and Natural Law", 75.

freedom is intricately tied to the problem of evil, these two problems in fact represent two interlinked aspects of the one problem of theodicy. To put it plainly, in a stream of time created and dominated by God's immediate and absolute power, how are we to deal with "the problem of the origins of evil"?[67]

5.3.1.1 *Redefining Freedom: Causality, Responsibility and Blame*

One of the problems scholastic theologians faced with Averroism was its appeal to the deterministic structure of the cosmos as grounds for freeing human beings from any moral responsibility for sin. After all, if human beings were compelled in their actions by the finite objects which surrounded them, then it would be unjustifiable to speak of personal responsibility. If the cosmos compelled evil in its created state, then this was at best the responsibility of the creator rather than the creature. Yet a need for personal justification and salvation required a basis upon human guilt for sin—a guilt often ensured by postulating a good creation and its fall via human *liberum arbitrium:*

> If God brought forth a good creation, and if there is no other evil divinity opposing God ... then evil must simply have its ultimate origins in one of God's—good!—creatures. And then this creature, in addition to its creaturely orientation toward God's will must have also had from the very beginning a possibility to oppose the will of God, to be able to decide for him but also against him. This is the weightiest theological argument for he acceptance of a *liberum arbitrium.*[68]

On the one hand, Pannenberg notes that the concept of a human free will has been repeatedly employed by the church in order to navigate its way through the problem of evil: preserving the essential goodness of God and God's good creation, yet maintaining a belief in human guilt.[69] But can such a free will be reconciled with the absolute power of a God

67 Pannenberg, "Christlicher Glaube und menschliche Freiheit", 274.

68 Ibid.

69 "[Die] Kirchenväter [hatten] im Kampf gegen die Gnosis auf das liberum arbitrium als den Ursprung des Bösen in der guten Schöpfung Gottes verwiesen"; ibid.

whose sovereignty rules over all time and all historical events?[70] This was a particularly awkward problem for Bradwardine due to his reliance on the *potentia Dei absoluta,* but it is equally vexatious for Pannenberg since he also argues that only the God who proves himself master over all is true, and that "to believe in one God means to believe that one power dominates all".[71] His declaration that "*every event* in which the future becomes finitely present must be understood as a contingent *act of God*"[72] and that this "is the only appropriate view of historical reality" only exacerbates his problem.[73] Given this position, it seems unlikely that Pannenberg would be able to avoid Bradwardine's error of making God the final responsible cause of evil as well as the immediate historical cause of human death and suffering.

Surprisingly, the young Pannenberg does not appear particularly disturbed by this problem or overly eager to avoid it. On the contrary, he argues that human beings have no reason to assume their own involvement in the appearance of sin in creation. Human beings, he argues, have adopted this stance purely out of doxological reasons. In our praise of the eternal creator we falsely attribute the guilt of sin in creation to ourselves.[74] Furthermore, Pannenberg finds it highly inappropriate for human beings even to raise "this impertinent question" [*vorwitzige Frage*] about the origins of evil.[75]

This leads to an awkward problem. Pannenberg seems to be liberating human beings from responsibility for evil in the world and laying this responsibility with the God who brings about all historical events. Yet he performs this liberation by rejecting the *liberum arbitrium* and thus pos-

70 Pannenberg, "Contingency and Natural Law", 114 n. 5.

71 Pannenberg, "Theology and the Kingdom of God", 55.

72 Ibid. 61, emphasis added.

73 Pannenberg, "Redemptive Event and History", 78.

74 "Es gehört zwar zum Wesen der Glaubenserkenntnis, nicht nur vor anderen Zeugnis abzulegen vom geschichtlichen Handeln Gottes, sondern auch von der Begegnung mit Gottes geschichtlichem Offenbarungshandeln zum Lobpreis seines ewigen Wesens aufzusteigen, so auch zum Lobpreis des Schöpfers und in eins damit zum Bekenntnis unserer Schuld als des Ursprungs des Bösen in Gottes guter Schöpfung"; Pannenberg, "Christliche Glaube und menschliche Freiheit", 275.

75 Ibid. 276.

tulating the absolute and determined slavery of humanity under the *potentia Dei absoluta.* Since all contingent events come directly from God, no human action can be historically efficacious either for "good" or for "evil". Such efficaciousness would return us to an understanding of time that moves from past to future and rejects the absolute contingency of historical events and God's role as the source of that contingency.

Thus Pannenberg needs a solution which simultaneously allows for the human guilt of sin yet also preserves the absolute power of God as sole historical subject. Pannenberg finds this solution by disconnecting the concept of human sinfulness from human action and linking it instead to an inherent, *psychological disposition* in human beings: namely "self-referentiality" [*Selbstbezogenheit*].[76] Pannenberg develops the details of this concept of dispositional (psychological) sin in his *Anthropology in Theological Perspective,* in an attempt to find a point of entry into the modern, non-theological discourse on anthropology. Interestingly, while the concept of self-referentiality evoking Augustine's statement on sin as an *incurvatus in se,* Pannenberg's treatment of sin and its relation to free will more directly reflects Luther's approach. In Luther's 1525 *De servo arbitrio,* his earlier and already opaque reading of Augustine now moves even further away from being identifiably Augustinian, as he argues that all things happen by absolute necessity,[77] and that God is the author of all evil human deeds.[78] McGrath notes Luther's significant divergence from Augustine, particularly on three points: (1) For Augustine, human beings are the authors of sin, whereas for Luther God is the author of sin; (2) Luther understands the slavery of human will as a consequence of human creatureliness rather than human sin; and (3) Luther directly argued for a doctrine of double predestination, whereas Augustine was unwilling to go down this route.[79] While it is clear that Pannenberg's approach stands more in line with this Lutheran tradition, McGrath stresses the direct "affinities" between Luther's work and Bradwardine's theology, noting that Luther even follows Bradwardine by leaving "no room whatsoever for contingency on the part of any created be-

76 Cf. ibid. 278.

77 WA 7.146.6–12; 18.615.12–17.

78 WA 7.144.34–145.4; 18.709.28–36. Cf. McGrath, *Iustitia Dei,* 229–30.

79 McGrath, *Iustitia Dei,* 230.

ing, including humans, whether they are sinners or not".[80] In this respect we find that both Luther and Pannenberg stand as Bradwardine's heirs.

Since for Pannenberg God is the subject of temporal and contingent events, Pannenberg prohibits himself from understanding sin as an active event—that is, a bodily action[81] we do with hands and "feet"[82] and eyes;[83] with mouths;[84] with rocks and stones;[85] an action we can commit against others,[86] which we can help others to avoid[87] and can ourselves avoid;[88] which can be counted and reckoned quantitatively;[89] or as a supra-individual power[90] which enslaves,[91] reigns over,[92] and binds the individual[93]. Such an active and consequential view of sin would detract from God's historical role as subject and from the *potentia Dei absoluta*. Rather than pursue the nuance in such biblical concepts of sin, Pannenberg instead appeals to social psychology (specifically in the work of G. H. Mead) since it can present an abstract cognitive anthropology satisfactorily distanced from bodily particularity.

Mead's social psychological theory presents advantages for Pannenberg since its concept of the self as constituted by outwardly-directed social processes over against the "sinful" recursiveness of the ego in self-referentiality[94] dissolves a concentration on human causal action and the

80 Ibid. 230 in general, also n. 63 with particular reference to WA 18.615.13–16.
81 Cf. Rom. 6:13, 7:23.
82 Cf. Mark 9:43–48.
83 Cf. 2 Peter 2:14.
84 Cf. Rev. 2:14.
85 Cf. Acts 7:59–60.
86 Cf. Mat. 6:14, 18:21; 1 Cor. 8:12; 2 Cor. 11:29; Jam. 2:9.
87 Cf. Mal. 2:6; 1 Tim. 5:20.
88 Cf. John 5:14, 8:11; Rom. 6:1; 1 Cor. 15:34; 2 Cor. 12:21; Eph. 4:26; Jam. 4:17; 1 John 2:1, 3:6.
89 Cf. Rom. 4:8, 5:20.
90 Cf. Eph. 6:12.
91 Cf. Rom. 6:16–22, 7:14, 25.
92 Cf. Rom. 5:21, 6:12.
93 Cf. Acts 8:23; Rom. 6:6.
94 Cf. Pannenberg, *Anthropology in Theological Perspective,* esp. 185–90. In his early work, Pannenberg traces this anthropology back to Duns Scotus (Pannenberg, "Person", 231–32). Pannenberg understands Duns Scotus as developing

concrete individual will as aspects of sin, and instead defines them both in terms of an existing social network. This abstracted social and cognitive anthropology thus allows Pannenberg to disconnect human sin (defined as regressive "self-referentiality")[95] from temporal or historical causality.

This at first seems counter-intuitive. Pannenberg confesses that if "sin is connected with the naturally transmitted concupiscential structure of behaviour, so that human beings are sinners from their birth and before any individual actions of theirs, then they can hardly be responsible for their sinful actions". But Pannenberg believes there is a way out if "responsibility" can be defined away from "the kind of responsibility that is *based on the principle of causality*".[96] This is what he then attempts.

Since for Pannenberg the only active and determinative force in creation must be the all-determining God, accepting the relationship between causal human action and sin would be a mistake which effectively limits the creative omnipotence of God. "The creative omnipotence of God excludes ... indeterminism as well as any inner-worldly determinism".[97] Pannenberg acknowledges that a dependence on human freedom, action

---

Richard of St. Victor's concept of the person as *existentia,* "als ein Bestehen (sistere), das von einem andern her (ek-) ist" (231), arguing that one can be a person in two ways: "in Verselbstung gegen Gott oder in gottgeöffneter Andacht" (232). Pannenberg thus moves to a reductionistic relational anthropology which defines the human being within a simplistic dialectic relation, first and foremost with God ("Der Mensch ist erst dadurch [Person], daß er Gott als [Person] sich gegenüber findet" [231]).

95 Cf. Pannenberg, *Anthropology in Theological Perspective,* 123.

96 Ibid. emphasis in original.

97 "Die schöpferische Allwirksamkeit Gottes schließt ... den Indeterminismus, so auch jeden innerweltlichen Determinismus aus"; Pannenberg, "Christliche Glaube und menschliche Freiheit"; ibid. 275. Note that Pannenberg still speaks here of a theoretical human freedom and independence, but only as an independence from the world: "Freiheit als Unabhängigkeit von der Welt", from "innerweltliche Bestimmungsgründe", from the "Determinismus innerweltliche Ursachen" and the "Totalitarismus innerweltliche Zwecke"; ibid. 276. But it is not clear what Pannenberg means by "freedom" in a situation where human beings are excluded from causal systems and subjected to the all-determining power of God.

and causality were indeed dominant ideas in the early, pre-Augustinian church fathers (and still had a role in Augustine's work),[98] but stresses that these "early Christian theologians did not give adequate thought to the price being paid for this solution": "aporias in the idea of a human freedom of choice".[99] Pannenberg's attempt to disconnect the concept of sin and responsibility from an understanding of active, or action-based causality seeks to overcome these problems since "only a concept of responsibility that is not one-sidedly linked with the concept of action but also extends to omissions and therefore to the person's own state can do justice to this approach".[100] By pointing at an unavoidable (yet anthropologically definitive) psychological disposition of recursiveness,[101] Pannenberg hopes to ensure that human beings still bear responsibility for sin despite the individual's (in)actions in creation: Need "responsibility

98 Cf. Clement of Alexandria, *Stromata* i.17; Origen, *De princ.* i, 3; Justin, *Apol.* 43, 4f.

99 Pannenberg, *Anthropology in Theological Perspective,* 110.

100 Ibid. 136.

101 Here we encounter an important methodological error in Pannenberg's *Anthropology,* based on a two-step procedure of reductionism and universalization. Pannenberg takes up Mead's work on self-referentiality and develops it in reference to a small sample group, in particular narrow circles of Germanic philosophical anthropology, in order to reduce the "universal" human being to this simplistic cognitive psychological definition. He then takes this reductionism and universalizes it, applying it to all humanity as his anthropological principal. It is clear that from Pannenberg's German context he is impressed with the interpretive idea of sin as radical *Selbstsucht,* egocentrism and self-referentiality. But can he prove here that he has not mistaken the cultural for the genetic? To support his argument that ego-centrism is the defining character of all human beings, he would need to include a far larger analysis of different cultural experiences of human life and society (including e.g. Middle Eastern systems of hospitality, as well as Asian and African models, etc.), rather than just presuming that the Western/Germanic individual is the universal exemplar for all peoples. In this dual movement of reductionism and universalization we have lost Schleiermacher and Kierkegaard's hard-won work on the importance and indispensability of the individual over against the homogenization and de-individualization evident in "universal" models; cf. e.g. F. D. E. Schleiermacher, *The Christian Faith,* ed. H. R. Mackintosh and J. S. Stewart, trans. D. M. Baillie et al. (Edinburgh: T & T Clark, 1928), §121.2.

[be] based on being the author or originator of something? When we say that someone 'accepts' or 'shirks' responsibility, we are speaking of actions that cannot be translated into judgements regarding causality".[102] Thus, Pannenberg's expectation is that human beings will heroically adopt "a concept of guilt that ... involves the 'acceptance' of responsibility by individuals for their own lives"[103] despite disconnecting such guilt and responsibility from one's own individual (*in*)*actions* and linking it instead to one's created state.

Pannenberg certainly shares here the challenge facing all post-Augustinian doctrines of original sin, namely to show how sin is an unavoidable reality for every human being, and yet still a just cause for responsibility, blame, punishment and salvation from God despite this unavoidability. But what exacerbates the problem for Pannenberg (in line with Luther and Bradwardine) is the rejection of all human contingency and causality which could ground the unavoidability of sin. This effectively fuses the problem of sin with theodicy and aggravates rather than eases the agitation that the problem of evil causes. All acts of sin and evil and suffering must come directly from the creative hand of the all-determining God, yet human beings (who have no causative power in history) are left responsible for this situation. The perceived injustice of this abstract model of sin must then be compounded by the concrete injustice of human legal systems which presume human causality.

We have seen that in his discussion with the natural sciences, Pannenberg suggests that time flows from future to past: each new moment of time is created directly by God and placed into temporality as our new present, and linked to the previous present (*Rückbindung*) through an act of God's faithfulness. Pannenberg uses this radical contingency to destroy the determinism of nature and do away with "natural laws". Pannenberg hopes that God's faithfulness in his continuing creation of time can be used to maintain the continuity of creation, accounting for the continuity of events in nature. But here we see that Pannenberg's approach has not only done away with the causality of *natural* laws but has also destroyed the causality of *human* laws. Jurisprudence and our systems of law are built on the notion of a continuity of events (of contingent act

102 Pannenberg, *Anthropology in Theological Perspective*, 112–13.
103 Ibid. 136.

and consequence), and ties this directly to the allocation of guilt, responsibility, blame and punishment. But Pannenberg's concept of divinely-created temporal moments and the lack of natural and human causation not only destroys a causative concept of sin but also the entire causative foundation of the legal system.

While jurisprudence is aware of the difficulties of reducing all views of guilt and responsibility to the individual alone, and acknowledges social psychology's sensitivity to issues of distributed responsibility and the contextualized attribution of responsibility,[104] this is far from eliminating all forms of causality from the legal system.[105] The legal concept of responsibility "refers to a decision about *liability for sanctions based on a rule*" which are themselves constructed by social expectations, but this already presupposes a basis on legal *causality:* "an actor is judged on the basis of causality (what *was* done) and expectation (what *should have been* done)".[106] Then in the subsequent attribution of blame, establishing the presence and limits of human *intent* (or the *mens rea*) as well as the legal *expectations* of others (especially in instances of negligence) are both important for legal evaluations.[107] The law must always negotiate this three-way matrix of causality, responsibility and blame. But if the causality of human action is eliminated from the temporal and historical system, and all causality of contingent acts is attributed to God, then the legal and ethical evaluation of human acts is eliminated along with it. Legally and ethically, it is difficult to conceive of human beings accept-

104 Cf. in legal studies e.g. J. Sanders and V. Lee Hamilton (eds.), *Handbook of Justice Research in Law* (New York: Kluwer Academic/Plenum, 2001), and in social psychology the foundational work on attribution theory by Fritz Heider, *The Psychology of Interpersonal Relations* (New York: Wiley, 1958).

105 When discussing responsibility attribution, sociologist V. Lee Hamilton suggested adopting a "motive grammar" (where accepting a situational role demand as normative leads to purposive attribution) and a "consequence grammar" (where rejecting the role demand is accompanied by causal attribution); V. Lee Hamilton, "Who Is Responsible? Toward a *Social* Psychology of Responsibility Attribution", *Social Psychology,* 41/4 (1978): 316–28.

106 Ibid. 316.

107 Cf. F. D. Fincham, J. M. Jaspers, "Attribution of Responsibility: From man the scientist to man as lawyer", *Advances in Experimental Social Psychology,* 13 (1980): 81–138.

ing responsibility and blame for temporal events when theologically (according to Pannenberg) both the causality and intent behind the event lie directly with God.[108] While Pannenberg focuses on the consequences of his concept of time for the natural *scientific* concepts of lawfulness and causality, less attention has been given to consequences for the *legal* concepts of lawfulness and causality.

Despite these efforts by Pannenberg to shift the guilt and responsibility of sin to human beings based on their dispositional state, the basic problem raised in Bradwardine's work is still in place. If only the God who proves himself master over all (including human history and time) is "true", if "to believe in one God means to believe that one power dominates all",[109] if "every event in which the future becomes finitely present must be understood as a contingent act of God",[110] and if this "is the only appropriate view of historical reality",[111] then despite Pannenberg's attempts to transfer the responsibility for sin and evil to human beings, God must remain the final and ultimate cause of both. *Each atomistic moment of historical time which is saturated with death, suffering and destruction must be interpreted as coming directly from the contingent will and creative hand of God, and actively desired by God for his creation.* If time is to be understood as a movement of sterile, independent temporal moments out of God's future and under the control of the sovereign rule of God, where the imperial *potentia Dei absoluta* determines all natural and human events, then there is no way in which theology can avoid making God the unmediated source of all evil, sin and suffering. A simple appeal to the recursiveness of human self-referentiality is unable to address the magnitude and concreteness of the problem that Pannenberg's concept of time raises. All instances of natural evil, from tsunami to disease, come directly from God's created future; all acts of sin all instances of moral evil, from individual acts of injustice to the genocide of zealous National

108 In God's newly created, absolutely contingent, undetermined moment of time, a person is murdered—but God leaves a human being holding the gun. This approach is even soteriologically awkward, since it would reverse the position of the church that in the cross, it is actually Christ who, though without blame, takes responsibility for human sin.

109 Pannenberg, "Theology and the Kingdom of God", 55.

110 Ibid. 61.

111 Pannenberg, "Redemptive Event and History", 78.

Socialism, are all events that stand beyond human causality and legal responsibility. They are all radically contingent events which encounter us from God's illimitable future: all represent a death and injustice that arrive from the active and immediate will of God.

This leads us to two further dangers.

5.3.1.2 *The Relativization of Suffering*

The first danger is an openness to the relativization of suffering by appealing to the hidden "truth of time" in God's "eternal present".[112] Such an argument seeks to relativize the suffering that occurs in history by arguing that the reality or "truth" of time can only be perceived from God's frame of reference. In other words, what appears to us as evil in the fragmentation of the temporal flow is in reality good once seen from God's eternity. This has two closely-linked "advantages". First, it allows for the destructive processes of the natural world to be equated with the acts of God. This then leads it, second, to bypass the problem of theodicy by claiming the inherent goodness of all those events that human beings otherwise experience as evil in history. This approach—which can be built upon Pannenberg's conception of time—functions by assuming that our encounters in time are somehow not trustworthy or dependable, and that an understanding of the "truth of time" can only be achieved when we see things from God's perspective.

To see how this works in practice, we can draw here on John F. Haught's "aesthetic cosmological principle" which takes this line, representing temporal suffering and death as features which promote and heighten the beauty of creation once seen from God's perspective. "Each occurrence in the world process, no matter how disagreeable it may be to our own moral sense, contributes novelty and contrast" to God's picture of the cosmos, even though it "may be locally disruptive at times".[113] By appealing to "God's feeling of the world",[114] Haught can then relativize what would otherwise be "irredeemable 'contradictions'", forming them into

112 Pannenberg, *What is Man?*, 77; Pannenberg, *Systematic Theology*, ii. 92; and cf. the discussion in the previous chapter.

113 John F. Haught, *God After Darwin: A Theology of Evolution* (Boulder, CO: Westview, 2000), 127.

114 Ibid.

"'contrasts' contributing intensity of beauty to [God's] 'wider vision'".[115] The only way that such an approach can function is by inserting a division between human experiences of time and God's true experience, between a limited human perspective and God's perfect frame of reference. The truth of time, the present, and reality is not that they are actually fallen and contain evil, but that the really real is located elsewhere, in God's view of all time.[116] Our difficulties with the existence of death and suffering in time is thus represented as the result of a skewed perspective away from the "significant beauty" of the "whole picture" that "comes to rest eternally in the empathy of God".[117] Once one has come to accept that God's perception of beauty "requires a balanced synthesis of the two polar extremes" one can transform all suffering, destruction, death and extinction into a mere "localized disharmony [which] can in the 'wider vision' become an intriguing shade that enhances the whole instead of a conflict that destroys it".[118] This type of theodicy can only function on the basis of a thoroughgoing dualism between experienced time and the "truth of time", between experienced reality and its true meaning—arrived at in some speculative method from within time. In Pannenberg's system, the danger becomes exponentially more hazardous since this dualism is attached to a rejection of historical causality. This allows not only for a relativization of natural evil but also for the ready sanctification of cultural and social evil, equating the sufferings and miseries of people on the margins of society and life not only with God's will but

115 Ibid. 128.

116 Ibid. 140. "An ethical outlook assesses everything in terms of human conceptions of good order, and so it remains baffled and often aggrieved by the chaotic consequences of the evolutionary openness to novelty. An aesthetic perspective, on the other hand, can accommodate the very contradictions that offend our moral sensitivity. It does so by allowing for a 'wider vision' and for the eventual transformation of clashes and chaos into a harmony of contrasts"; ibid., 130–1.

117 Ibid. 130. "[Critics] cite evolution's 'wastefulness' and its 'cruelty' or 'indifference' toward the weak. ... Such a judgement, however, can easily overleap the fact that for the most part the universe sustains life instead of crushing it. Furthermore, even the perishing of individual organisms—including ourselves—seems to be consistent with the ongoing adventure of a larger life-process"; ibid. 130–1.

118 Ibid. 130.

also with God's conception of beauty and truth—all while divinizing the eager propaganda of cultural dominance.

What is missing in Pannenberg's work, as with Haught, is a deeply authentic appreciation of a Christian *theologia crucis,* a faith in a God who personally suffers the pains and deaths of time's own fallenness, rather than standing at a safe distance behind them. Theology must not resort to trite assertions of the mere "appearance" of the brokenness of time; cosmic and historical time must not simply be relativized as non-genuine forms of time, rather they exist as genuinely fallen, authentically destructive and truly death-dealing to human personhood and all creation, and must therefore look towards healing and fulfilment in God's redemption of all creation.

Despite Pannenberg's dependence upon a concept of the controlled "eternal present" of an "all-determining God" as "the truth of time", and while his theology of time could serve as the basis for such a relativization of suffering, to his credit he does not pursue that argument. Indeed, Pannenberg agrees with Barth in his condemnation of those who would "reinterpret reality by trivializing its shadow side".[119] "Suffering, guilt, and tears cry out for a real overcoming of evil", and no "purely theoretical theodicy" can achieve such an end.[120] On the contrary, if "the objection is to be met, then it will be met only by a real overcoming of evil and suffering such as Christian eschatology hopes for in faith in the resurrection of the dead".[121]

Yet while the intention is there, and while Pannenberg is aware of the dangers, his simple *theologia gloriae* is so overwhelmingly inflated with a concentration on the power and sovereignty of God that it leaves no room for a complementary *theologia crucis* or for any truly nuanced perspective on the problem or existence of suffering, either within human history or within the divine life.

5.3.1.3 *The Instrumentalization of Time*

In the discussion of God's faithfulness in the process of *Rückbindung,* Pannenberg noted that "forms of uniformity in natural occurrences

119 Pannenberg, *Systematic Theology,* ii. 164.

120 Ibid.

121 Ibid.

[can be represented] as an expression of God's faithfulness, the identity of which is based not on a natural unchangeability but on adhering to the decisions once made in each case".[122] The key point Pannenberg makes here is that the natural world maintains its coherency in time not by physical principles of inertia or the continuance or development of physical states from the past into the present, since "every event in which the future becomes finitely present must be understood as a contingent act of God".[123] Rather it is only as God's "backward-reaching incorporation of the contingently new into what has been, but not the reverse, as a predetermining mastery and its effect, [that] ... the primary connection of history [can] be conceived without losing its contingency".[124] Thus it is God who creates each new contingent moment of time, and actively links each new present to the immediate past. Yet because of God's absolute power and the radical contingency of all events, God is neither constrained nor obliged to ensure that each new and contingent moment of time should in fact connect with the last. God's creation hangs upon God's continuing desire to sustain and maintain it. And in the option to derail creation, to inflict death and annihilation upon his creatures either by natural, social or historical forces, God expresses his omnipotent judgement over creation, over human society and the church. In this regard, God need not just use natural disasters but can also use the forces of Islam, the confessional divisions of the church, the two World Wars, the spectre of socialism, or any other radical contingency that may yet arise from God's open future to punish his creation.[125] Each new moment of

122 Pannenberg, "Contingency and Natural Law", 109.

123 Ibid. 61.

124 Pannenberg, "Redemptive Event and History", 76.

125 The "universalism of the Byzantine empire itself was perverted into exclusivism, especially by imperial insistence on dogmatic uniformity. The judgement of God on that perversion of the Christian vocation was specifically spelled out in the rise of Islam"; Pannenberg, "Election and History", 102. "The dogmatic exclusivism of the Christian sense of chosenness was continued ... on the basis of medieval clericalism. God's judgement on that deformation of his church occurred in the divisions of the church. ... These led to the period of confessional wars"; ibid. 103. "[God's] Judgement on the pride of nationalism ... came to expression most clearly in the self-destructive outburst of antagonisms of European nationalism in World War I that put an end to the dominating role of Europe

time is radically contingent and not open to natural causality, therefore all new instances of evil and suffering that occur in time come directly from the will of God. Such an instrumentalization of time affirms time's destructive power but also links such destructiveness to the omnipotent will of God and the freedom and contingency of God's future. In this way time is removed from the realm of fallen creatures. The brokenness and tragedy of time is embraced and transformed into a weapon of divine vengeance upon creation, rather than representing time as itself an aspect of creation in need of the salvific, redeeming action of God.[126]

5.4 Fields of Force

With the foundations and frameworks for Pannenberg's concept of time in place, we then gain a better sensitivity for instances in his later work where his conception has changed and where it has remained largely the same. If we turn to Pannenberg's *Systematic Theology,* we notice a softening in his stance. Pannenberg offers explicit though still somewhat ambivalent expositions on the issues of human action and the problem of theodicy. His early alignment with Bradwardine on the *potentia Dei absoluta* as the ground of all contingent natural and human events gives way slightly to allow for a limited independence of creatures over against God, and even for limited efficacious action in determining future states—though there is no wholesale change in his approach. In his discussion of Moltmann's use of thermodynamics, Pannenberg concedes:

---

in the world. It meant that the divine vocation that was perceived earlier in experiences of national chosenness, had been forfeited by nationalistic self-glorification. That judgement became definitive with World War II"; ibid. 104. The "ambiguities of [the] secular conception [of liberty] deprived the idea of liberty of its holistic meaning to such an extent that judgement on the societies based on that principle is already looming on the horizon. This judgement obviously appears in the rise of socialism"; ibid. 105.

126 There is an underlying inconsistency in Pannenberg's work here. On the one hand, God (in his omnipotence and in the freedom of his future) is the source of every natural and historical event. Yet in the end human beings do seem able to act with causality in history: at least by prompting God's judgement upon themselves.

> there may be a reversal of the direction of time [i.e. a reversal which makes time run from *past to future,* rather than from future to past!] as present events restrict the field of possibility. This partial determination of the future by the past and the present, which in classical mechanics and electromagnetics is close to complete determinism, need not be judged negatively from a theological standpoint, even as a temporal inversion of the structure of events vis-à-vis its basis in micro-occurrence. For it is a condition of the continuity of natural events and is thus an expression of the faithfulness of the Creator in his will to bring forth independent creaturely forms. This inversion of the temporal structure and the mode of working of natural forces may thus be regarded as a sign of their creaturely independence in relation to the creative dynamic of the divine Spirit from which they have emerged.[127]

This passage is remarkable for several reasons and displays significant points where Pannenberg has developed over his earlier work. While he still envisages the sequence of temporal events running from future to present and past (proceeding from God's eternity[128]), Pannenberg does acknowledge the possibility not only of a reversal of this order but that the future field of possibility may even be limited by "*independent* creaturely forms" in the present, leading to a "*partial* determination of the future". As a result, Pannenberg opens a small door for an independent basis of creaturehood, one which would allow creatures an efficacious role in time and history.

But despite the significance of this move, it is still far from a total revision of his earlier thought. This small change still represents an "inversion" of the proper order of time, and "independence" does not mean self-determination, but rather the creature's "duration" in time and "distinction" (i.e. removal) from the "eternity" of God which is its proper "origin".[129] There

---

127 Pannenberg, *Systematic Theology,* ii. 108.

128 Ibid. 92.

129 "The future toward which creative forms move in the duration of their existence has for them an ambivalent face. On the one hand, for the preservation, development and consummation of their nature, they are referred to a future over which they have little or no control, while on the other hand, since they are finite, the future threatens to end and dissolve their independent form. It is precisely because of their creaturely independence, and the related distinction from their origin in creation, that creatures are exposed to the fate of the dissolution of their form"; ibid. 96.

can be no true self-determination because the future represents a realm over which creatures "have little or no control",[130] the realm of God's eternity. Pannenberg is also clear that such a "temporal inversion in the structure of natural forces and their operation causes them to become ungodly and demonic forces ... when they close themselves against the future of God, the kingdom of his possibilities, and thus become closed systems".[131] It is an odd anthropomorphism here to speak of natural forces and fields of operation—systems of molecules and energy—"closing themselves" against God's future. But this leads us to another change in his approach: Pannenberg shifts his model to that of field theory, transposing his ideas into a new metaphorical framework. Yet despite the change in metaphor, the core conceptions see no significant change.

In Pannenberg's somewhat awkward adoption of the metaphors of field theory,[132] he picks up an approach pioneered by T. F. Torrance who sought to link the function of the Spirit in creation to physical descriptions of nature.[133] Whether this move is an advance over Pannenberg's previous efforts is open to debate, especially as Torrance's success with this model was also notably limited.[134] What is clear is that Pannenberg was impressed by the theory's chances of accounting for "the effective presence of God in every single phenomenon" via a pneumatological connection.[135] Beyond the model of fields of force, physics had concentrated on concepts of force as linked directly with particular bodies or masses—a problem for a theology wishing to attribute force in crea-

130 Ibid.

131 Ibid. 108.

132 The following discussion draws upon Max Jammer, *Einstein and Religion: Physics and Theology* (Princeton: PUP, 2002), 153–266.

133 Cf. T. F. Torrance, *Space, Time and Incarnation* (Oxford: OUP, 1969). "We must credit T. F. Torrance with being the first to recognize these links and plead for theology's adoption of the field concept"; Pannenberg, *Systematic Theology*, ii. 82.

134 Cf. e.g. Polkinghorne's concern that Torrance's work comes "dangerously close to Newton's equation of absolute space with the sensorium of God"; J. Polkinghorne, *Reason and Reality: The Relationship between Science and Theology* (Philadelphia: Trinity, 1991), 93.

135 Wolfhart Pannenberg, "The Doctrine of Creation and Modern Science", *Zygon*, 23 (1988): 3–21.

tion to a non-substantial divinity.[136] Yet in his eagerness, Pannenberg too quickly links the field theory of modern physics to the pre-Socratic teachings of Anaximenes, and then appeals to Max Jammer's argument in the *Historisches Wörterbuch der Philosophie*[137] to claim "the modern field concept" as a direct development of "the Stoic doctrine of divine *pneuma*".[138] He then links this hypothesis (following Rüsch[139]) to a Christian doctrine of the Spirit: "This Stoic doctrine influenced not only the work of Philo but also what early Christian theology had to say about the working of the divine Spirit in creation".[140]

Yet there are difficulties with the transposition, and Pannenberg's adoption of field theory has attracted widespread critique from physicists who see it as a misinterpretation of the physical theory.[141] As Wicken has argued,[142] in pre-relativity physics the field of space was linked to an ether filling the void (by which force could then be transmitted); but with the advent of relativity physics, the Newtonian "void" ceased to exist, replaced by the matter of the four-dimensional matrix.[143] Thus space itself became a field. "If all the matter were removed from the universe,

136 "If all force comes from bodies or masses, then and only then the understanding of natural events is definitively cut loose from any connection with the concept of God, since God cannot be understood as body"; Pannenberg, *Systematic Theology*, ii. 80.

137 Max Jammer, "Feld—Feldtheorie", in J. Ritter (ed.), *Historisches Wörterbuch der Philosophie* (Basel: Schwabe, 1972), ii. 923–7, esp. 925.

138 Pannenberg, *Systematic Theology*, ii. 81.

139 Theodor Rüsch, *Die Entstehung der Lehre vom Heiligen Geist bei Ignatius von Antiochia, Theophilus von Antiochia and Irenäus von Lyon* (Zürich: Zwingli, 1952).

140 Pannenberg, *Systematic Theology*, ii. 81.

141 It is not my intention to provide a detailed overview of the entire body of critique here. As a start, cf. e.g. Polkinghorne, *Belief in God in an Age of Science*, 82; J. S. Wicken, "Theology and Science in the Evolving Cosmos: A Need for Dialogue", *Zygon*, 23 (1988): 45–55; H.-D. Mutschler, "Schöpfungstheologie und physikalischer Feldbegriff bei Wolfhart Pannenberg", *Theologie und Philosophie*, 70 (1995): 543–58.

142 Wicken, "Theology and Science in the Evolving Cosmos", 45–55.

143 This is a difficulty Ted Peters also runs into with his view that "a force field is an independent reality *prior* to the body. Body and mass become *secondary* phenomena"; Ted Peters, "Pannenberg on Theology and Natural Science", 13.

there would be no field.... Space and matter have coevolved, and are relationally constituted by each other. They have no identities apart from each other".[144] Therefore, if Pannenberg's intention is to find an entry point by which God (as a non-substantial entity) can enter into physical theory, then field theory is hardly sufficient for the task. Either God is conceived of (as creator) as beyond the plane of created entities which compose the field, or God himself is reduced to a field—"ubiquitous", "without intentional structure"[145]—which then places God not only on the level of created objects but makes the concept of God superfluous (since the physical space-field already exists and is theoretically sufficient). Wicken is harsh in his critique: "If we want to use the word 'field' in science–theology discourse, let us do so in some way commensurate with their understandings in physics. Talking about 'spirit' as 'energy' and granting it by implication the status of physical law runs dangerously close to usurping the hard-won denotative language of science for physicalizing theology".[146] But the harshest critique is levelled by Polkinghorne, who dismisses Pannenberg's theory as "quaint".[147]

From the perspective of the sciences, Pannenberg's use of field theory appears highly problematic. But does the concept help him theologically? Does his use of field theory alter his stance on causality, contingency, and the independence of creatures (subject to the *potentia Dei absoluta*) or help overcome the problems of theodicy? In fact at first it does seem that Pannenberg is willing to concede a certain amount of independence for creation and its creatures, noting that the "regularity of events in elementary processes is the basic presupposition for the development of permanent forms. ... The uniformity of events according to law is thus a condition of creaturely independence"![148] Yet the underlying foundation remains unchanged, with the "uniformity of natural occurrence" still a determined "expression of God's faithfulness and constancy in *his*

144 Wicken, "Theology and Science in the Evolving Cosmos", 52.

145 Michael Welker, "Introduction", in id. (ed.), *The Work of The Spirit: Pneumatology and Pentecostalism* (Grand Rapids, MI: Eerdmans, 2006), pp. ix–xviii at xi.

146 Wicken, "Theology and Science in the Evolving Cosmos", 48.

147 Polkinghorne, *Belief in God in an Age of Science,* 82.

148 Pannenberg, *Systematic Theology,* ii. 72.

activity as Creator and Sustainer".[149] An independent uniformity and coherence of natural events still cannot be thought of without the direct creative activity of God. In part, this expected conclusion stems from Pannenberg's use of a doctrine of divinely determined continuous creation which focusses upon direct divine involvement, upon God's constant creation and sustenance of the cosmos at each temporal moment.[150] In this respect, there seems to be a degree of affinity with process theology, and some have seen Pannenberg's later work (with its focus on field theory and the creative "persuasion" of God) as a gradual appreciation of Whitehead's work.[151] However, Pannenberg himself quickly draws the line at what he sees as difficulties with process theology's restriction of the Creator's omnipotence:

> By restricting the power of God, the theological adherents of Whitehead's process philosophy seem at a first glance to be able to offer a more illuminating answer to the experience of wickedness and evil than the Christian doctrine of creation. In truth however, this teaching leads to the result that the creature does not depend on God alone but on other powers, so that it cannot rationally put full trust in God alone for the overcoming of evil in the world.[152]

The absolute position of God as sole source of determinative power is not a point Pannenberg is ready to negotiate. The stress on God's *potentia absoluta,* God's "unlimited", "infinite", "omnipresent" and "eternal" "power over all things" remains.[153] It seems that Pannenberg is aware of the persistent dangers of his approach, and attempts to avoid them, arguing that such omnipotent rule "is not the excessive omnipotence of tyranny", which would not allow for the independent existence of crea-

149 Ibid., emphasis added.

150 Cf. ibid. 38; also 34, 40, 42, 49, 51, 122–3, 130, 142.

151 Cf. Lou Ann Trost, "Non-interventionist Divine Action: Robert Russell, Wolfhart Pannenberg, and the Freedom of the (Natural) World", in Ted Peters and Nathan Hallanger (eds.), *God's Action in Nature's World: Essays in Honour of Robert John Russell* (Aldershot: Ashgate, 2006), 205–216, esp. 211–12.

152 Pannenberg, *Systematic Theology,* ii. 16.

153 Cf. Pannenberg, *Systematic Theology,* i. 415–16.

tures.[154] Yet Pannenberg's system draws him back to the same conclusions: these creatures in their existence cannot possess any ability for efficacious or self-determining action. The unlimited "power of God [must have] no precondition outside itself ... bringing forth that over which it has power".[155] To say that God's "omnipotence cannot be totally opposed to [creatures]" since "as Creator, God wills [their] existence" is true at a fundamental ontic level.[156] But in the praxis and flow of time, the will of creatures is still utterly subsumed into the determining power of their creator. For Pannenberg, this simply must be so if we are not to neglect the close connection of "the doctrine of God's omnipotence ... with the act of creation and the resultant lordship (*potestas*) of God over his creation".[157] Taking his lead from the Socinian doctrine of God, Pannenberg asserts that the "supreme being is called God in virtue of his all-embracing power and lordship (*potestatem et imperium*). Without these he would not be God once anything exists outside himself".[158]

Thus in the final analysis, the mechanism of God's continuing creation of each temporal moment (whether by a process of *Rückbindung* or within a field of force) makes little significant difference if the omnipresence and omniscience of the all-powerful and all-determining God still serve to create (and thus validate) each new moment of time and history. While such an approach can lead to the *affirmation* of creation (despite its fallenness) as stemming from the hand of God, it provides little help for *confronting* the brokenness and suffering of time, or for providing theological support for the battles of the oppressed.[159]

154 Ibid. 416.

155 Ibid.

156 Ibid.

157 Ibid. 417.

158 Ibid. 417 n. 173.

159 In this vein, Pannenberg dismisses a range of contextual theologies—such as feminist theology and liberation theology—seeing them as examples of the excessive assimilation of theology into modernity's culture of secularisation; cf. Wolfhart Pannenberg, *Christianity in a Secularized World* (London: SCM, 1988), 47–56. In response, Pannenberg's approach has been dismissed e.g. by representatives in the South African context for being overly "authoritarian" (cf. e.g. Cornél du Toit, "Fin de siécle and the development of a post-secular religion: intimations from the TRC", *Religion and Theology*, 4/1 [1997], available

In the end, Pannenberg side-steps the issue by arguing that regardless of the debate surrounding the free will or independence of creatures, ultimate responsibility for suffering and evil still lie with God,[160] and that God's response to that evil is seen in the (still unfulfilled) promise of Christ's resurrection.[161] Thus Pannenberg argues that our stance toward the problems of evil and suffering cannot an the active and causal alteration of present time, but rather a patient waiting for the final eschatological redemption and consummation of the world under the hand of God:

> A belief in creation that does not go hand in hand with hope of eschatological victory over the reality of evil and sin ... can give no answer to the question of theodicy. An answer is possible only when we see that creation is linked to the divine work of the reconciling and redeeming of the world. If creation comes to completion only with the world's reconciliation and redemption, then the creator is allied to us in the battle to overcome evil and to reduce and heal suffering in the world. Only the eschatological consummation of the world can definitively demonstrate the righteousness of God, and therefore his deity, in the work of creation.[162]

For Pannenberg, once the eschatological fulfilment of the divine promise occurs, God will have proven his deity and his final victory over suffering and death. We exist now in that period which precedes the fulfilment of the promise made in the resurrection of Christ.[163] Yet once again,

---

online from the University of South Africa <http://www.unisa.ac.za/default.asp?Cmd=ViewContent &ContentID=7368>, accessed 20 June 2007), or coming under suspicion for being used by Christian supporters of the apartheid movement. My thanks to Prof. Michaelis Christoffel Dippenaar for this insight.

160 Pannenberg, *Systematic Theology,* ii. 166.

161 Wolfhart Pannenberg, "Dogmatic Theses on the Doctrine of Revelation", 142.

162 Pannenberg, *Systematic Theology,* ii. 173.

163 This expansion of time between promise and fulfilment constitutes one of Pannenberg's earliest definitions of history and brings him quite close to Moltmann's theology: "History arises because God makes promises and fulfils these promises. History is event so suspended in tension between promise and fulfilment that through the promise it is irreversibly pointed toward the goal of future fulfilment"; Pannenberg, "Redemptive Event and History", 18, 31; and Pannenberg, *Systematic Theology,* i. 251.

the uncertainty involved in the fulfilment of God's promise,[164] and its relocation purely into an eschatological realm can lead to a trite relativisation of suffering as well as a nullification of lament[165] and of intercessory prayers which seek indeed a pre-eschatological, historical liberation from suffering.

5.5 Summary: *"Gott gänzlich unterworfen"*

In the previous chapter, we examined the concept of time which Pannenberg developed over the course of his theological career, noting in

164 Yet in line with the broader thrust of his theology (with its focus on the illimitable power of God), Pannenberg argues that God may indeed make promises, but is under no obligation to fulfil these promises in line with our expectations. God has the freedom to fulfil the content of these promises in ways contrary to how they were initially promised or understood (cf. e.g. Pannenberg, "Redemptive Event and History", 31). Thus there is a fundamental uncertainty regarding the contents of what tradition claims God has promised, which can ultimately serve to reduce faith in the "concrete particularity" of God's historical promises. If the content of the promise can be changed under appeal to divine power, to what extent can the promise remain meaningful as a promise to those who depend upon it? In rough terms, each promise has (a) a central content and (b) a range of peripheral elements related to this central content. One might change the peripheral elements or fail to define them in exact terms but retain the central content; or define the peripheral elements while leaving the central content vague. One might promise a child a dog, but not specify colour or breed or sex or even the timing of the promise's fulfilment, leaving these points either assumed or simply undefined. Yet in some cases a change of the peripheral elements may affect the central content of the promise to such a degree that the promised object is no longer desirable: the dog may be dead. While still within the bounds of the "letter" of the original promise, in a promise between unequal partners, it must be questioned whether such an approach that alters the "spirit" of the promise is just or even helpful since it renders the content of that which is hoped for so vague that it negates the "concreteness" of the historical promise.

165 For more detail on the difficulties associated with the concept of lament in such a "passive" eschatology, see Stephen Lakkis, "'Have you any right to be angry?' Lament as a metric of socio-political and theological context", in Eva Harasta, Brian Brock (eds.), *Evoking Lament: A Theological Discussion* (Edinburgh: T & T Clark, 2009), 168–82.

particular the specific impulses and motives which shaped his work, such as a concern with the natural determinism of the sciences and its link to a developmental (and deterministic) understanding of the way the cosmos functions; the assimilation of this method into the historical sciences via the principle of historical analogy; and Pannenberg's concern regarding the negative impact this confluence of features had on the historical claims of the Christian faith (namely, that God should be able to work and reveal himself in history and to bring about novel events without analogy and in apparent contradiction to the so-called laws of nature). Pannenberg also saw belief in the historical resurrection of Christ (as a novel and contingent event without analogy) called into question.

In response, Pannenberg focused on the absolute power of God to bring about any event from God's own open future, which is free from natural determinism by natural "laws". Because of God's omnipotence, God can express his sovereign rule by bringing into existence each new moment of time and linking these moments to the past, either coherently (in his faithfulness) or incoherently (in divine judgement).

In this chapter, I suggested that the essential underlying structures of Pannenberg's position are adopted from the period of the medieval theological conflict with deterministic Averroism. Indeed, we could argue that Pannenberg's significant achievement as a medieval scholar was to notice the similarities between the medieval context and the current debate with the natural sciences, and to creatively reapply that medieval solution in the modern world.

By noting the startling affinities between Pannenberg's work and the theology of Thomas Bradwardine (in turn transferred through Luther), this chapter showed how Pannenberg revives this medieval response to Averroism, which relied strongly on the *potentia Dei absoluta,* on the overwhelming power, rule and sovereignty of God. We saw how medieval Christian theologians faced in Averroism an influential, academic system which argued for natural determinism,[166] threatened the sovereignty of God,[167] limited (if not effectively excluded) the unmediated ac-

166 Dolnikowski, *Thomas Bradwardine,* 190.

167 Pannenberg, "Bradwardine and the Pelagians", 359.

tivity of God,[168] and allowed for a naturalistic course of development.[169] Further complicating matters, early scholastic realism had begun to focus on the discovery of universal and eternal truths, neglecting historical particularity.[170] In reaction, critics of this approach attempted to pull the focus back from ideals onto particulars,[171] and in doing so appealed to the *potentia Dei absoluta* as the basis both for all contingency[172] and for coherency in the natural world.[173] It was Bradwardine in particular who would extend the principle of the *potentia Dei absoluta* to its logical end in its "absolute" jurisdiction over all nature and all humanity.[174] By appealing to God's "eternal present" this control could also be expressed over all time.[175] The result was an emphasis on God's sovereignty and power in history and nature, and the particularity and contingency of all temporal events.

The parallels to Pannenberg's analysis of the modern theological discourse with the natural sciences are striking: theologians face a dominant scientific system which argues for a natural form of determinism either via inertia[176] or the invariable space-time manifold,[177] oppressed by "a world of eternal and unbreakable laws which [govern the individual's] own life as well as the realm of nature".[178] The unmediated activity of God in the natural world has been surrendered to a naturalistic form of historical and cosmic development lacking contingency or novelty.[179] An idealistic focus on absolute and universal truths (natural laws) has led not only to a rejection of novelty, but even to the exclusion of basic data when it fails to fit into the predictive systems of natural laws[180]—an

168 Ibid. 357; Dolnikowski, *Thomas Bradwardine,* 165.
169 Pannenberg, "Bradwardine and the Pelagians", 357.
170 Placher, *A History of Christian Theology,* 162–7.
171 Ibid. 163.
172 Pannenberg, "Archbishop Thomas Bradwardine", 174.
173 Pannenberg, "Bradwardine and the Pelagians", 359.
174 Ibid. 357.
175 Dolnikowski, *Thomas Bradwardine,* 184.
176 Pannenberg, "Contingency and Natural Law", 109–13.
177 Stannard, "God in and beyond Space and Time", 114.
178 Pannenberg, "The Biblical Understanding of Reality", 17.
179 Eddington, *The Nature of the Physical World,* 286.
180 Ibid.

approach which has spread to the historical sciences in the form of the analogical method, ruling out the historical resurrection of Christ.[181]

To respond to these challenges, this chapter argued that Pannenberg practiced a neo-Bradwardianism, resurrecting the theological concept of the *potentia Dei absoluta* and stressing God's unconditioned power over all natural, historical, and human events. God's absolute power is expressed by bringing into existence and maintaining each radically contingent, naturally disconnected and historically particular event[182] and connecting them all in a temporal flow from future to present to past.[183] God's absolute power as well as his faithfulness is expressed in the coherent order of these contingent events.[184] God's absolute power ensures that all temporal events, natural and human, spring from the will of God, since both nature and history are unable to secure their own development in time.[185] God's absolute power is expressed in his ability to link these events into any (in)coherency either in accordance with his faithfulness or judgement.[186] God's absolute power is expressed in his absolute control over all time from the realm of his omnipotent and eternal present.[187] And the absolute power of God is expressed in the absolute sovereignty of his being and rule as the *all-determining reality*.[188] For all these reasons, Pannenberg can stress so univocally: God's "being is his rule".[189] Thus how can anything challenge God's absolute power without simultaneously detracting from his being? Mirroring Bradwardine's argument against both deterministic Averroism and so-called "Pelagian" nominalism, Pannenberg appeals to the *potentia Dei absoluta* to subdue and (literally) overpower what he sees as the deterministic natural and historical sciences, using God's absolute power to defend the idea of God as an absolutist ruler who intimately controls both nature and the events of human history.[190]

181 Pannenberg, "Redemptive Event and History", 43–50.
182 Pannenberg, "Theology and the Kingdom of God", 59.
183 Pannenberg, *Systematic Theology*, ii. 92.
184 Pannenberg, "Heilsgeschehen und Geschichte (Fortsetzung und Schluß)", 285.
185 Pannenberg, "Contingency and Natural Law", 83.
186 Pannenberg, "Heilsgeschehen und Geschichte (Fortsetzung und Schluß)", 285.
187 Pannenberg, *Systematic Theology*, ii. 92.
188 Pannenberg, "Contingency and Natural Law", 114 n. 5.
189 Pannenberg, "Theology and the Kingdom of God", 55.
190 Pannenberg, "Contingency and Natural Law", 75.

The practical advantage of the *potentia Dei absoluta* is that it is a catch-all, self-justifying theological claim which, as such, can hardly be argued against.[191] If absolute power is predicated of God and linked with God's being, then nothing can detract from God's being or God's power. Furthermore, since all other power is less than absolute, God's power necessarily trumps all other claims. The principles of historical analogy or the lessons of experience fall before it, accused of having a narrow methodology; the absolute power of God dismantles so-called "laws of nature", rendering them mere "descriptions" of God's past presents. By what right could science or historiography (or other theological views)

191 Physics certainly has its own equally trite, catch-all argument, namely a simplistic concept of the infinite universe (or infinite universes). An infinite universe can conveniently operate as a justification for an infinite set of possibilities. This line of argument appears periodically in cosmology (especially regarding the possibility of extraterrestrial life); see e.g. G. F. R. Ellis and G. B. Brundrit, "Life in an Infinite Universe", *Quarterly Journal of the Royal Astronomical Society,* 20 (1979): 37–41; or J. Lunine, "The Frequency of Planetary Systems in the Galaxy", in Ben Zuckerman and Michael Hart (eds.), *Extraterrestrials: Where are They?* (Cambridge: CUP, 1995), 192–214. Apart from being enormously speculative and metaphysically wasteful, Ian Hacking has also critiqued this approach for expressing the inverse gambler's fallacy: that the occurrence of an improbable event leads to the assumption that it occurred within a series of many similar tries (the outcomes of which were more probable). This leads to a logical circularity: the mere occurrence of an improbable event appears to validate the theory of a multiplicity of "tries" or even a (necessary) infinity of tries depending on the degree of "improbability" of the original event. Cf. Ian Hacking, "The Inverse Gambler's Fallacy: The Argument from Design. The Anthropic Principle Applied to Wheeler Universes", *Mind* NS, 96/383 (Jul., 1987): 331–40. The argument is extended in Milan M. Cirkovic, "Is Many Likelier Than Few? A Critical Assessment of the Self-Indication Assumption", *Epistemologia,* 27 (2004): 265–98, esp. 289–90. In physics, Paul Davies dismisses an appeal to the infinite, noting that "Invoking an infinite number of other universes just to explain the apparent contrivances of the one we see is pretty drastic, and in stark conflict with Occam's razor.... I think it's much more satisfactory from a scientific point of view to try understand why things are the way they are in *this* universe and not to invent invisible universes to do the job"; Paul Davies, *Are We Alone? Philosophical Implications of the Discovery of Extraterrestrial Life* (New York: Basic, 1995), 121.

set limits upon God's power, or to dare limit his freedom? By outdoing and outbidding all other claims to power, the *potentia Dei absoluta* claims necessarily to overcome all external critique.[192]

However, several serious problems do arise from use of the *potentia Dei absoluta* and from Pannenberg's specific adaptation of it. First, it has its roots in a period of political absolutism, where the concept of absolute power had analogues in both state and church. We could identify this origin as the likely cause of the allergic response the concept produces in postmoderns, especially given our pluralist (and especially postcolonialist) sensitivities. Yet the major difficulty facing a doctrine of absolute divine power is by far the problem of evil. To his credit, Pannenberg is well aware of the problem, as was Bradwardine. Bradwardine's overwhelming focus on the *potentia Dei absoluta* in both nature and in human events led, despite his efforts, to an unavoidable divine determinism and the final attribution of sin and evil and suffering to God.[193] But Pannenberg fares just as badly, unable to overcome the same ending to which the doctrine pushes him. Three phases were noted in the development of Pannenberg's work on this problem.

In his early work, Pannenberg did not appear overly concerned about the problem of evil. Following in Bradwardine's and, more directly, Luther's footsteps, Pannenberg was satisfied to dismiss human responsibility for evil as a doxological presumption.[194]

In his later *Anthropology,* he displays a more nuanced engagement with the issue. Many classical theodicies preserved God's goodness through an appeal to human freedom, action and historical (developmental) causality, thus also ensuring the role of human beings in sin and their need for human redemption.[195] Since for Pannenberg God is the sole subject of natural and human history, he runs the risk of absolving human beings of the need for redemption and salvation. His attempted

192 Note the affinities here with Thomas Aquinas' fourth proof (*ex gradu*) for the existence of God; cf. Thomas Aquinas, *Summa Theologiæ,* 1a, Q2, a3. If power exists to varying degrees in creation, one might assume the perfection of power in absolute form in God.

193 Cf. Pannenberg, "Christlicher Glaube und menschliche Freiheit", 274 for discussion of humanity's inability to resist God.

194 Ibid. 275.

195 Pannenberg, *Anthropology in Theological Perspective,* 110.

solution was to disconnect the concept of responsibility from that of action and link it instead to an underlying state or disposition[196]: namely an unavoidable human recursiveness or self-referentiality which opposed the fulfilment of human destiny in exocentricity.[197] Despite the unavoidable state of human recursiveness and the absence of effective causality in time and nature,[198] Pannenberg then encourages human beings to accept the guilt and responsibility "for their own lives".[199]

This chapter argued that a range of problems arise from such an approach. The social- and cognitive-psychological method which Pannenberg adopts is highly reductionistic, discounts the bodily nature of human existence as well as the extensive biblical testimony to the active bodily dimensions of sin.[200] Indeed, overall coherency with the biblical traditions and texts represents a significant problem for Pannenberg since the biblical witnesses do not represent the human being in narrow cognitive-psychological terms, or portray sin as the simple recursiveness of the individual human psyche. Nor can Pannenberg's disconnection of intent, causality, responsibility and blame be easily integrated into an ethical or legal system—rather it serves to destroy our understandings of causative legality and coherent ethical behaviour. Regardless of this shift into a psychological anthropology, the problem of God's role as the originator of evil and historical suffering is still insurmountable. Each atomistic moment of historical time which is saturated with death, suffering and destruction must be interpreted as directly willed and coming from the hand of the omnipotent God, actively desired by God for his creation, irrespective of non-causal human egocentrism. Indeed, by shifting the stress away from human actions, Pannenberg ensures that *all* natural and human experiences of suffering must now be attributed in an unmediated way to God.

---

196 Ibid. 136. Following a similar approach taken by Luther, who in turn expressed Bradwardine's influence; cf. McGrath, *Iustitia Dei*, 230.

197 Cf. Pannenberg, *Anthropology in Theological Perspective*, 123, 185–90.

198 Cf. ibid. 112–13, 123.

199 Cf. ibid. 136.

200 These bodily aspects certainly do not exhaust our descriptions and understandings of sin, but neither can they simply be dismissed. We need to find ways to incorporate the richness of these biblical traditions into our conceptions of sin.

This led to two further problems: the relativization of suffering (i.e. the claim that suffering and evil are actually "good" if viewed from the perspective of God's eternal present, thus negating the brokenness of time and its need for healing), and the instrumentalization of time as a tool of God's judgement over creation.

Finally, when we come to Pannenberg's *Systematic Theology,* we see a late reappraisal of his model and a shift to the concept of fields of force. If we overlook Pannenberg's misunderstanding of the physical concept of fields of force (which would render the metaphor impracticable), in theory this move displays promise, especially since it could allow for efficacious human action in the present and a "partial" self-determination of creation.[201] But in practice Pannenberg's advances are limited. Creaturely independence represents an "ungodly and demonic" "inversion" of the proper temporal structure[202] and leads us to death;[203] time still runs from future to past and has its truth in God's eternal present;[204] and the coherence and uniformity of natural occurrences is still dependent upon God's faithfulness[205] and direct, continuing creation.[206] Most significantly, God has lost none of his absolute omnipotence, maintaining his "unlimited", "infinite", "omnipresent" and "eternal" "power over all things".[207] Even the promising link between process theory and field theory is rejected because Pannenberg sees it shifting determinative power away from God.[208] In the final analysis, God's being is not seen in Jesus, not seen in the crucified Son of Man, not seen in the One who is Emmanuel, but rather in the *theologia gloriae* of a philosophical God whose perfect being is seen in his absolute and imperial rule over all.[209]

201 Pannenberg, *Systematic Theology,* ii. 72, 108.

202 Ibid.

203 Ibid. 96.

204 Ibid. 92.

205 Ibid. 72.

206 Ibid. 34, 38, 40, 42, 49, 51, 122–3, 130, 142.

207 Cf. Pannenberg, *Systematic Theology,* i. 415–16.

208 Pannenberg, *Systematic Theology,* ii. 16.

209 Cf. also Pannenberg, *Systematic Theology,* i. 417 n. 173: the "supreme being is called God in virtue of his all-embracing power and lordship (*potestatem et imperium*). Without these he would not be God once anything exists outside himself".

What then should we make of Pannenberg's work on time? To answer that question (and before looking at some final points of critique in this volume's conclusion), perhaps it's best to draw on Pannenberg's own words in his criticism of Bradwardine. In his work on time, Pannenberg

> simply replaces a cosmically mediated determination, operating through the powers of the cosmos, with an unmediated determination through God himself. Even the human being, while free from compulsion through finite things, is in each of life's movements completely subjugated to God [*Gott gänzlich unterworfen*]. ... there remains no room for creaturely independence. ... God [is] the decisive, driving force in all creaturely events.[210]

So that the thesis that Pannenberg "did not advocate a theological determinism ... [is a thesis] one can hardly follow. While it is clear that he attempted to avoid the deterministic consequences of his doctrine of God",[211] nevertheless "the force of his systematic approach thrust him to deterministic consequences against his own will".[212] "The principle pushes him even to trace sin back to God as its author, and to understand God's will as the ultimate cause even of damnation".[213] And "although this trend stands in tension with [his] intentions, as his various attempts to stave off the logical consequences of his approach reveal",[214] Pannenberg's theological motives "do nothing to change the fact that the systematic-logical structure of his approach has a deterministic character",[215] and leaves us with little reason to hope in the overthrow of suffering, or in the defeat of the brokenness of time, either now or in the future.

210 Pannenberg, "Bradwardine and the Pelagians", 357.
211 Pannenberg, "Archbishop Thomas Bradwardine", 175.
212 Ibid.
213 Pannenberg, "Bradwardine and the Pelagians", 357.
214 Pannenberg, "Archbishop Thomas Bradwardine", 175.
215 Ibid.

CHAPTER SIX

CONCLUSION: RULE BY LAW, RULE OF LAW

In his work on time, Pannenberg aimed at developing a new theory of time which could satisfactorily account for theological and scientific experiences of the world. He sought to develop a temporal theory that could preserve the freedom of God to be involved directly in time, bringing about new and unforeseeable occurrences in accordance with God's will, while simultaneously accounting for the empirical experiences of natural regularity, continuity, and developmental causation.

Perhaps surprisingly, we found that the concept of time that Pannenberg developed bore several similarities to a classical scientific understanding of time: they are both linear, they are both objective, and they are both deterministic in character. The core difference in Pannenberg's approach is that he swaps out the causative agent of that determinism, using a theological doctrine of the illimitable and absolute power of God to outbid natural determination, trumping it with an even greater deterministic power. Rather than attacking the underlying determinism upon which physical systems of temporality depended, Pannenberg embraces it, but places it under the rule of a new master: he swaps de-personalized rule by natural laws for the direct deterministic will of the omnipotent God.

But in following a medieval, philosophical conception of God as all-determining reality, has Pannenberg wandered too far away from the Christian God revealed in the cross of Jesus Christ? Can a wilfully-dominating and deterministic system that has no room for a *theologia crucis* adequately express the truth of God's reality in and with time? In this chapter, I will give a final overview of the scientific discussion on time as well as Pannenberg's response, and argue that classical physics and Pannenberg alike put forwards conceptions of time based upon a reductionistic and undifferentiated view of lawfulness. To make this evident, I will draw on the distinction in jurisprudence between "rule *by* law" and the "rule *of* law". We will then be able to see that both the classical sciences and Wolfhart Pannenberg seek to institute temporal systems based on rule *by* law, and I will suggest that theology is better served by developing a contrasting system based on the rule *of* law. In specifically theological terms, our theological goal should be to replace a dominant and deterministic "rule *by* God" with the open and supportive "rule *of* God".

6.1 Rule by Natural Laws

The great accomplishment of Newtonian physics was its regularity, which is based on calculable laws of nature. Whereas for Ockham, "God was still indispensable in the explanation of the physical world because without God no finite reality could persist", Newton's work on inertia and his laws of the conservation of energy replaced "the dependence of physical reality on God's activity of continuous creation with the idea of *self*-preservation".[1] Just as natural selection would later lead to the biological sciences' own "coming of age" (providing the basis for self-sufficiency in the development of life), Newton's groundbreaking principle of inertia formed "the basis of all laws of nature"[2] and marked physics' own age of maturity.[3] From that time on, the motion of the universe was disconnected from the mysterious forces of a divine hand.[4] The great ad-

1 Pannenberg, "Theological Questions to Scientists", 19, emphasis added.

2 Pannenberg, "Contingency and Natural Law", 109.

3 Inertia "played ... a decisive role for making nature independent toward its divine origin"; ibid.

4 Consciousness of this "coming of age" is expressed by Newton when he notes in his preface to the *Principia* that: "the ancients ... made great account of the science of Mechanics in the investigation of natural things; [but] the moderns laying aside substantial forms and occult qualities, have endeavoured to subject the phenomena of nature to the laws of mathematics"; Isaac Newton, *The Mathematical Principles of Natural Philosophy*, trans. Andrew Motte (3rd. edn.; London: Motte, 1729), n.p. In his own preface to the second edition, Newton's colleague Roger Cotes was not so reserved, clearly emphasising this view that this new school of scientists were breaking away from their predecessors who "tell us nothing". Thus "laying aside that useless heap of words, [we] thought to employ [our] pains to better purpose"; Roger Cotes, "The Preface of Mr. Roger Cotes, To the Second Edition of this Work, so far as it relates to the

vances achieved by Newtonian mechanics were based upon an underlying conviction regarding the lawfulness of motion: that motion was self-regulating, that inertial bodies would retain their states until acted upon by other forces, and that these interactions were immensely predictable since they were governed by an observable, though above all calculable and replicable set of mechanical laws. As a lawful system, Newtonian mechanics also promoted a temporal ontology which pictured the physical universe "unfolding" in strict accordance with natural law and within the framework of absolute time. It was on this basis that the Enlightenment could replace the free and personally involved "savior God of medieval and Reformation theology" with "the clockmaker God of eighteenth-century deism".[5] Unlike theology, which dealt with the contingent acts of a divine personality, Newton's standardized, mathematical formulas seemed able to reveal the clockmaker's own laws for the operation of the universe. Thus, so long as one had the sufficient raw data for completing the formulas, one could predict the future state of physical systems, including the broader universe. Given complete knowledge of the past and present, classical physicists believed they could predict the future completely.[6]

The result of this advance was that the future no longer represented an impenetrable realm of uncertainty or contingency which needed to be approached and controlled through the comforting rites of the cult, but was in principle wholly predictable and knowable. Likewise the present, as the future moment of the immediate past, was also sterilized of any contingency, since it was merely the lawful and mechanistic extension of that physical system which existed immediately prior to it. We must not underestimate the importance of this predictive ability in controlling and taming the contingency and uncertainty of an unknown future, as it forms the basis of an immense system of belief and trust in the natural sciences: "What is the ground for so much ardent faith in the orthodox hypothesis that physical phenomena rest ultimately on a scheme of com-

---

Invention and Discoveries herein contained", in Isaac Newton, *The Mathematical Principles of Natural Philosophy,* trans. Andrew Motte (3rd. edn.; London: Motte, 1729), n.p.

5 Lindberg, "Introduction", 12.

6 Cf. Eddington, *The Nature of the Physical World,* 228.

pletely deterministic laws? ... [It is because the] principal laws of Nature which have been discovered are apparently of this deterministic type, and these have furnished the great triumphs of physical prediction".[7] It is for this reason that Newton presents such a significant marker for the beginning of modern science. By expropriating "control of the future" from the church and placing it within the comprehension of the lawful and thus predictable sciences, his work furthered the scientific revolution's rapid move away from the exclusive authority of the church. Where cultic and religious systems seemed to have failed, Newtonian physics was confident that in principle it had finally removed contingency and uncertainty from the physical world. While contingents still appeared in practice, such contingency was merely the illusion that arose from insufficient knowledge or calculation errors. With patience and diligence this apparent contingency could also be eliminated.[8]

At the turn of the twentieth century, Poincaré's work on the "principle of relativity", and Einstein's fulfilment of the idea, continued this tradition in the physical sciences. While in some respects it was true that Einstein's work represented "the downfall of classical physics",[9] Einstein still stood upon Newton's shoulders and defended the strict, determining lawfulness of the universe and its associated predictive power over the future.[10]

Yet despite this stress on predictability, the universe of Newtonian physics still displayed aspects of openness. "Absolute space" and "absolute time" formed an all-encompassing receptacle in which "absolute motion" ("the translation of a body from one absolute place into another") still occurred.[11] The predictability of Newtonian mechanics arose from applying the laws of motion to those bodies shifting about in absolute space. But with Minkowski's unification of space and time, the concept of an open future in which objects could move came to an end. Rather than individual three-dimensional objects shifting about in the

7 Ibid. 286.

8 "It was, of course, recognized that we were only acquainted with part of the structure of this causal scheme, but it was the settled aim of theoretical physics to discover the whole"; ibid.

9 Ibid. 13.

10 Cf. Ibid. 142–3.

11 Newton, *The Mathematical Principles of Natural Philosophy*, 9–10.

container of time, Minkowski offered an understanding of the universe as a collection of invariant four-dimensional objects which together defined the block universe: a four-dimensional unity lacking any external dimensions; an invariant solid that remains constant and is already existent. "[F]our-dimensional space-time does not change. Something can only change *in time.* But space-time is not 'in time'. *All* of time is [already] there ... *past, present, and future.* It's static".[12]

Rather than objects popping in and out of existence from past to future or being translated from one space to the next, the invariant nature of Minkowski space-time demands that objects be considered in their totality, in their extension as complete four-dimensional entities. Minkowski advanced on Einstein's work by guaranteeing matter a radical permanence across time, rather than merely in the "present".[13] The "past" was no longer non-existent and the "future" was no longer an empty space waiting to be filled by the motion of objects out of the present. The Minkowski universe presented stable objects which exist and extend as a four-dimensional whole; and these objects collectively were seen as coterminous *with* four-dimensional space-time, rather than stratified within space and time.[14] To think of an object only in its present state without including its entire existence as a four-dimensional solid extending from the absolute past into the absolute future was an abstraction which failed to do justice to the temporal distension of that object.

Newtonian cosmology had worked with the idea of an open space in which entities changed and advanced. Newton's laws of motion prescribed the ways in which those objects progressed, and the movement of objects within space formed the basis of a concept of time. In contrast, in the Minkowski space-time manifold we deal simply with the temporal and spatial extension of static entities. At the level of the underlying philosophy of relativity, the laws of relativity therefore no longer *prescribe* motion but rather *describe* the extension and curvature of non-moving, invariant, four-dimensional solids. Here, lawfulness took on a new dimension of meaning: no longer prescriptive of action, the law serves now to validate the block universe as it is. The difference is both in quality and

---

12 Stannard, "God in and beyond Space and Time", 113.

13 Weinert, *The Scientist as Philosopher,* 128.

14 Eddington, *The Nature of the Physical World,* 56.

in degree. Newton's laws, in their prescriptive nature, still worked with the introduction of impressed forces to shift objects out of their inertial states and determine their motion in the future.[15] However, Minkowski space-time does away with talk of determining the future in a prescriptive sense because the future is already complete in its extended four-dimensionality and cannot be changed any more than the past. From the perspective of the space-time manifold, the Renaissance Man who actively shaped the world around him becomes as much a fiction as the interventionist God acting through a universal *Heilsgeschichte*. The lawfulness of the natural world is rendered absolute, not because mechanics can predict how objects will move in the future, but because the future state of those objects is already existent and cannot be changed. Expressed in theological terms, we shift from a Newtonian perspective where creation might be imagined as the establishment of an initial state "in the beginning" which then grows and develops (in a lawful way) into the future, to creation of the Minkowski universe as the coming into being of these four-dimensional solids in their entirety, *en bloc*. This makes the creation (the space-time manifold) a static one, not in the sense that there is no movement, but in that no variance can be introduced, neither into these four-dimensional solids nor into the laws which describe them.

6.2 Rule by Divine Decree

The reaction to Newton's work on the strict lawfulness of the natural world varied. David Hume was representative of those who saw the work as signalling the final elimination of divine interference in history due to the power and pre-eminent rule of natural law.[16] Others (including

15 "An impress'd force is an action exerted upon a body, in order to change its state, either of rest, or of moving uniformly forward in a right line"; Newton, *The Mathematical Principles of Natural Philosophy*, 3.

16 "A miracle is a violation of the laws of nature; and as a firm and unalterable experience has established these laws, the proof against a miracle, from the very nature of the fact, is as entire as any argument from experience can possibly be imagined. Why is it more than probable, that all men must die; that lead cannot, of itself, remain suspended in the air; that fire consumes wood, and is extinguished by water; unless it be, that these events are found agreeable to the

Newton[17] and Cotes[18]) were not as eager to preclude the role of God from the world, seeing natural laws as providing precise descriptions of the way in which God interacted with nature. Unfortunately, such a "lawful" position threatened to remove any aspect of the personal from an understanding of God, transforming God into simply another form of universal law. As Pannenberg notes, a personal conception of God's involvement in the world presupposes a certain degree of contingency, of personality, making it awkward if not meaningless to speak of necessary and universal laws as "personal decisions". Thus a third group (of which Pannenberg is representative) perceives the strict lawfulness of the natural world as a threat to God's divine power and sovereignty and as a danger to God's divine right at each point in time to decide in which ways he sustains creation. Here, both the existence and the movement of objects must stress the radical contingency of God's decisions in creation: rather than expressing lawful self-sufficiency, the occurrence and movement of objects in time express divine decision, divine will, and divine purpose.[19] From this perspective, we can understand the clash between the strict rule of natural laws and the rule by divine decision as an argument about power and the location of that power—whether it is contained in a natural structure and system or whether it is based on the decisions, motivations and involvement of a personal being[20]: "inertia" is

laws of nature, and there is required a violation of these laws, or in other words, a miracle to prevent them? Nothing is esteemed a miracle, if it ever happen in the common course of nature"; Hume, *An Enquiry Concerning Human Understanding,* x.1.

17 "Newton, however, believed that God has a continuing active role in the physical world. ... Newton saw the laws of nature as a continuing expression of God's sovereignty and will, acting in a regular way"; Barbour, *Religion and Science,* 22–3.

18 The "best and most valuable fruit of [Newton's] philosophy, [is to incite] the more profoundly to reverence and adore the great Maker and Lord of all. He must be blind who from the most wise and excellent contrivances of things cannot see the infinite Wisdom and Goodness of their Almighty Creator, and he must be mad and senseless, who refuses to acknowledge them"; Cotes, "The Preface of Mr. Roger Cotes", n.p.

19 Cf. Pannenberg, "Dogmatic Theses on the Doctrine of Revelation", 125–31.

20 Pannenberg, "Contingency and Natural Law", 109.

legalistic, whereas "decision" bears the stamp of a personal power, not of a mere structure of laws.[21]

In the science and theology discourse, Wolfhart Pannenberg has been one of the most significant proponents for such a reclamation of divine power, based on his concerns that natural laws functioned as a direct limitation on the jurisdiction of God's personal and absolute power: "Christian apologetics, having accepted the new basis of natural philosophy provided by the principle of inertia, was now left to the unfortunate strategy of looking for gaps in the continuity of natural processes if it wanted to preserve certain occasions for divine interference in the natural world".[22] For Pannenberg, the principle of inertia and the "laws of nature" which have been built upon it have usurped the need for God to be involved in each temporal event, for God to make "decisions" regarding the status of events "in each case",[23] not just in the present but also in the future. Thus at worst, the lawfulness of nature exorcises God from the universe; at best, the system of law-bound development which arises from the precepts of classical physics relegates God and God's activity to the past (as a "first cause" or as creator only "at the beginning"[24]) doing away with the concept of divine power in the course of time, and content to deal with the developmental growth and progression of time and the cosmos. Time becomes the simple expression of the unfolding of predetermined tendencies already contained within an embedded, internal seed.[25] In such a system, Pannenberg fears there is no room for the power of God, or for the novel events which God may want to bring about, and certainly no room for a historical resurrection from the dead.[26] It is for these reasons that Pannenberg sees an urgent revision of the principle of inertia and its effects on time as not only necessary but as one of the most crucial issues to be addressed in the dialogue between science and theology.[27]

---

21 Ibid. 113.

22 Pannenberg, "Theological Questions to Scientists", 20.

23 Pannenberg, "Contingency and Natural Law", 109.

24 Ibid. 72.

25 Cf. Pannenberg, "Redemptive Event and History", 42f.

26 Pannenberg, "Heilsgeschehen und Geschichte (Fortsetzung und Schluß)", 266–7 n. 22.

27 Pannenberg, "Theological Questions to Scientists", 20.

As his own response to that dialogue, Pannenberg develops a theory of time which specifically aims to reinstate the power of God over creation. He begins by reiterating his position that if history and nature run within their own deterministic frameworks (based on the rule of natural laws and developmental progression), then not only does "the notion of a personal power behind those events [become] untenable"[28] but such an inertial "notion of self-movement" would mean that the resultant entelechy could never be the grounds for hope in a novel future: the future would simply unfold itself out of an "already present germ" and thus can never be the basis of something "really new".[29] For these reasons, Pannenberg's counter-thesis is that it cannot be the past which determines the present and future; rather it must be the future which determines the present and past.[30]

Pannenberg supported this thesis by his early development of a concept of time which runs from the future to the past, and which is held together by the concept of divine "backward linkage" [*Rückbindung*].[31] In essence, each moment of time that appears to us as the present does not unfold out of the past (subject to rule by natural laws), but is rather a contingent moment created by God in the freedom of his future and which is then placed into the stream of time by God, whose faithfulness and omnipotence ensures that each of these discrete moments of time link back and cohere to one another.[32] Several points are vital here: In such a system there simply is no lawfully-constrained, predetermining past which could place limits on God's power and will to act in creation, since we are dealing with purely disconnected and contingent events.[33] For Pannenberg, it is God's "future which lets go of itself to bring into being our present",[34] rather than the present being a direct and lawful extension of the immediate past.[35] Since there is here no "predetermining mastery of the past", and no basis for the operation of natural laws the absolute sovereignty and rule of God are preserved.

28 Pannenberg, "Theology and the Kingdom of God", 58.
29 Pannenberg, "Appearance as the Arrival of the Future", 139.
30 Pannenberg, "Theology and the Kingdom of God", 59.
31 Pannenberg, "Heilsgeschehen und Geschichte (Fortsetzung und Schluß)", 285.
32 Ibid.; and cf. also Pannenberg, *Systematic Theology,* ii. 72.
33 Pannenberg, "Heilsgeschehen und Geschichte (Fortsetzung und Schluß)", 285.
34 Pannenberg, "Theology and the Kingdom of God", 59.
35 Pannenberg, "Contingency and Natural Law", 106.

The power and rule of God is central to Pannenberg's understanding both of the nature of God and the concept of time. His assertions in this regard—that if our image of God "does not comprehend *the power dominating everything,* ... [it is] not a true concept of God"[36]; that "Only the God who proves himself master over all is true"[37]; and that "The deity of God is his rule"[38]—form the constant refrain that runs through the corpus of his theology. Even in his later work, where Pannenberg explores the power of God in time via field theory, the absolute position of God as sole source of determinative power—his "unlimited", "infinite", "omnipresent" and "eternal" "power over all things"[39]—is not a point Pannenberg is willing to negotiate on. Whether it is in nature or history, God's creatures fail to possess authentic ability for efficacious action or self-determination since the unlimited "power of God [must have] no precondition outside itself", it "bring[s] forth that over which it has power".[40] Finally, for Pannenberg God's power and rule over all must be absolute since the "supreme being is called God in virtue of his all-embracing power and lordship (*potestatem et imperium*). Without these he would not be God once anything exists outside himself".[41]

In the end, the mechanism of God's direct and unmediated continuing creation of each temporal moment, either in a process of *Rückbindung* or or within the metaphor of a field of force, makes little difference to Pannenberg's final goal: namely, to stress the *potentia Dei absoluta* over all events, and God's freedom to direct creation over against all so-called laws of nature.

In this respect, we saw that Pannenberg's response to the natural sciences is not so very different from that proposed by Thomas Bradwardine in response to deterministic Averroism and the concerns of "Pelagian" scholarship in the fourteenth century, which were then re-expressed through Martin Luther. This is perfectly fitting for Pannenberg as a Lutheran theologian with a specialization in medieval theology. Just as Brad-

36 Pannenberg, "Theology and the Kingdom of God", 60, emphasis added.

37 Ibid. 55.

38 Ibid.

39 Cf. Pannenberg, *Systematic Theology,* i. 415–16.

40 Ibid.

41 Ibid. 417 n. 173.

wardine swapped out divine determination over natural determination in his debate with Averroism,[42] Pannenberg replaces determinism by inertia and the natural laws of physics with determination by the direct and unmediated creative action of God as omnipotent and all-determining reality.

The concept of time Pannenberg offers as a result if this approach is not so different from the scientific model he attempts to critique. In short, both systems can be described as advocating a "rule *by* law" approach. In the natural sciences, the absolute and universal laws of nature direct each event in a deterministic manner, specifying their orderly progression in time within the bounds of an underlying and unchanging mathematical framework. In Pannenberg's work, the absolute and universal law is God's will and power which bring into being each new temporal event, specifying their orderly linear progression within the bounds of a divine determinism based on the concept of absolute and all-embracing might and lordship. What this leaves us with is a metaphor of the clash of two imperialistic powers: both systems remain linear, both systems remain objective, and both systems remain deterministic—and both systems claim absolute and unqualified rule. The essential difference is that Pannenberg's concept of time presents us with an absolutist "rule by divine decree" rather than a de-personalized "rule by natural laws". Thus Pannenberg's answer to the absolute power represented by natural laws is not an attack on the underlying concepts of determinism, but rather to trump them with the concept of an even more powerful determinant in the form of the absolute power of an all-determining God whose very being is his absolutist rule. Science may talk of laws of nature, but for Pannenberg these crumble before the absolute rule of God.

Unfortunately, Pannenberg's stress on the absolute power of God drives him to the same unavoidable difficulties that Bradwardine and Luther faced before him. The advantage to a scientific focus on natural determinism is that the sciences need not concern themselves with defending the concept of a loving God. The problem for Pannenberg is that any theological system that radically stresses God's being and role as all-determining reality inevitably splinters against the rocks of historical suffering, evil, and sin. If the absolute power of God truly determines all historical events in a direct manner, then there is no way to distance the

42 Cf. Pannenberg, "Bradwardine and the Pelagians", 357.

horrors of life, the savagery of nature, and the atrocities of history from the express will of God. Pannenberg attempts to steer away from the danger with a reliance on God's promise and future fulfilment,[43] but is finally unable to overcome the momentum of his own system. Pannenberg's assimilation of scientific "rule by law" under the omnipotence of God pushes him to redefine both the human being and sin in shallow and reductionistic psychological terms, to relativize and endorse suffering and evil in history, and to instrumentalize time as the tool of God's judgement rather than itself being a fallen part of creation in need of redemption. Pannenberg's failure to deal adequately with the evils and sufferings of history reflects the absence in his theological approach of a thoroughgoing *theologia crucis* or a significant engagement with the sufferings and death of God in history. Today, the voices of countless victims around the globe who cry out in pain and anguish from their temporal situations of torture, torment and affliction remain the final and insurmountable critique of any simplistic and bourgeois Western *theologia gloriae* that imagines an imperialistic, all-determining reality with its *potentia absoluta* can pass as the God of love revealed in the cross of Christ.

6.3 Rule by Law versus Rule of Law

The problem facing both Pannenberg and the deterministic sciences in their use of "rule by law" models is this very concept of lawfulness and "lawful rule". The concept of law is a thoroughly polyvalent one which has a wide range of meanings not only in jurisprudence but also in the natural sciences, in economics, sociology, as well as theology. Furthermore, "law" maintains this polyvalent character even *within* each of these disciplines. Within jurisprudence, for example, the discourse on the so-called "rule of law" has long delved into the various aspects of meaning behind the term. I will argue here that the distinctions and insights which have arisen in that legal discourse can help us further in our own discussion on lawful progression in time.

In 1995, Chinese law professor Li Shuguang critiqued the Chinese state's approach to the law by noting that "Chinese leaders want rule by

43 Pannenberg, *Systematic Theology*, ii. 173.

law, not rule of law". The difference, Li noted, is that "under the rule of law, the law is preeminent and can serve as a check against the abuse of power. Under rule by law, the law can serve as a mere tool for a government that suppresses [people] in a legalistic fashion".[44] Two important points arise here: First, with the concept of the rule of law comes an underlying assumption regarding the establishment of a legal system or framework within which the entire community exists—ruler and subject, government and citizen. All are subject to the influence of the legal framework because all exist within it. Second, the framework of the rule of law ensures a degree of individual freedom and self-determination. While the law still serves to structure society, it also guarantees individuals certain freedoms to operate within that structure—bounded by the limits of the laws, of course, and subject to judgement when infringing upon those limits—but otherwise supported by the clear lines of social inter-relationality (lines solidified and made clear for example by precedent and the traditions of common law). In this sense, the rule of law serves a *supportive* function for the life of a society and the individuals who comprise it. The law surrounds our lives and offers a protective and defined space in which we can live and move. In this way the law can fulfil its promise, becoming that which "provides the setting in which the human spirit can develop in freedom and diversity".[45]

In contrast, rule by law seeks the opposite: the creation (by sovereigns) of a comprehensive framework of laws within which only the "governed" (rather than the "governors") exist—and where only the governed are subject to its influence. Rule by law *instrumentalizes* the law, transforming it into a tool to oppress those under it and determine the intricacies and minutiae of everyday life.[46] Rather than a system which provides support and the freedom for self-determination, the law then

44 Quoted in Tamanaha, "The Rule of Law for Everyone?", 101.

45 *Declaration on Democratic Values,* G7 Economic Summit, London, June 8 1984, Article 2.

46 Chen Jianfu notes that in the Chinese context, sought-after reform of the rule by law legal system is often translated into making the system more *efficient,* rather than more *egalitarian:* i.e. not changing the instrumentalized nature of the law, but merely extending its effectivity and its reach over more aspects of individual and social life; Chen Jianfu, "Market Economy and the Internationalisation of Civil and Commercial Law in the People's Republic of China", in

is used to direct the individual's each action towards the ends pre-established by those who rule by it. As an instrumentalization of the law, it is essentially characterized by the confluence of sovereign power and control. "[R]ule by law is distinguishable from rule of law in that the former rejects the central premise of rule of law that law is to impose meaningful limits on even the highest government officials"[47] who then use the law to exercise determinative or directional control over its citizens, so that "a rule by law system aims at a much higher degree of thought control".[48]

The distinction between rule of law and rule by law is not absolute. There are moderate and extreme versions of rule by law regimes, and it may be difficult to distinguish the boundary line between a moderate rule by law system and a corrupted rule of law system.[49] The British jurisprudential heritage and the cultural presuppositions which infuse the concept of "rule of law" are also open to critique.[50] But developing a sensitivity to the distinction between rule of law and rule by law can bring us

---

Kanishka Jayasuriya (ed.) *Law, Capitalism and Power in Asia* (London: Routledge, 1999), 69–94.

47 Randall Peerenboom, "Competing Conceptions of Rule of Law in China", in id. (ed.), *Asian Discourses of Rule of Law: Theories and Implementation of Rule of Law in Twelve Asian Countries, France and the U.S.* (London: Routledge, 2004), 109–41 at 120.

48 Ibid. 121.

49 Peerenboom warns that this difficulty also exists towards the other extreme: the Chinese "legal system during the Mao era, particularly during the Cultural Revolution, was a good example of an extreme version [of rule by law], to the point where at times it hardly could be described as even a rule by law legal system, which after all implies some form of law-based order"; ibid. 120.

50 Deborah Cao has noted that the Chinese *fazhi* (法治) can be translated as either "rule of law" or "rule by law", since at a linguistic level the Chinese language does not differentiate between the two. Furthermore, "the rule of law is regarded as presupposing political or economic structures of liberal democracy" and other contents such as basic human rights and the welfare state—a sociopolitical basis which rules it out from universal adoption. Deborah Cao, *Chinese Law: A Language Perspective* (Aldershot: Ashgate, 2004), 35–55, esp. 35–6. In this respect, the rule of law is "a cultural achievement of universal significance", but one that needs to have this cultural dimension highlighted before declaring it a universal and "unqualified human good"; E. P. Thompson, *Whigs and Hunters: The Origin of the Black Act* (London: Allen Lane, 1975), 265–6.

further in the analysis of concepts of time. Seen from this perspective, we can argue that relativity and the inertial systems of classical physics provide a system of rule by natural laws, where each part of the natural world is strictly determined by a structure of laws which specify and prescribe the exact manner of its causal progression. While there is no authority outside or beyond these laws (the laws are simply assumed as given) the universe exists within the framework of these given laws and is completely directed by them. Pannenberg rejects this scientific rule by natural law and replaces it instead with his own rule by divine law, where law functions in the sense of a divine decree (that God's word and will are his law). Pannenberg's model runs identically to the scientific approach he is critiquing, but whereas natural determinative laws are simply assumed (since for physics there can be no transcendent authority "outside" the universe), it is precisely this transcendent authority which Pannenberg gives us in the form of the all-determining God. Wielding his *potentia absoluta,* God exists outside the lawful system and instrumentalizes it in his deterministic control of time and creation.

If we extend this analogy, we might ask if there are physical and theological options for an alternate rule *of* law model. With quantum theory, fundamental physics has already begun to make this shift away from a system of strict temporal and physical causality. In the late 1920s, at the watershed period of this development, Eddington had already noted that physics was now confronted with a new epistemological system,[51] one which "no longer pledged [physics] to a scheme or deterministic law".[52] Admittedly, the changeover from strict causality to a probabilistic system (which arose from Heisenberg's work on quantum indeterminacy) only represents a first small step in this direction,[53] but the consequences are significant. Because the majority "epistemology of science presupposes a deterministic scheme" in its view of nature, these new physical models are demanding deep changes in the way we perceive nature and natural

51 Eddington, *The Nature of the Physical World,* 221–4, 292–4.

52 Ibid. 283.

53 While it is true at a theoretical and philosophical level that quantum theory deals with "probabilities" rather "certainties", it remains to be seen whether in practice quantum physics will simply continue to use "certainty" as a shorthand term for "very highly probable".

knowledge.[54] Einstein's own urgent appeal "to restore unison between physical reality and the profoundest characteristic of Newton's teaching—strict causality" highlights the degree of anxiety and concern that this shift produced.[55] This apprehensiveness comes from the disruption of the link between information and causality in the task of predicting the future. Unlike classical physics which envisioned complete knowledge of the future from complete knowledge of the past, Heisenberg's work and the so-called "new physics" suggested that "the future is a combination of the causal influences of the past together with unpredictable elements—unpredictable not merely because it is impracticable to obtain the data of prediction, but because no data connected causally with our experience exist".[56] The result of this disconnection is to open up the future, "to say that the future cannot be anticipated", except with a certain degree of inaccuracy.[57] "The classical view ... amounts to disguising a piece of the unknown future as an unknowable element of the present. Classical physics foists a deterministic scheme on us by a trick; it smuggles the unknown future into the present, trusting that we shall not press an inquiry as to whether it has become any more knowable that way".[58] If we apply the metaphor of rule of law here, then the idea is not that quantum physics is not lawful, but rather that it presents an overarching system of laws which provide for a degree of openness for the movement of atomic particles, rather than attempting to prescribe that motion in concrete, deterministic and causal terms.[59] The impact and importance

54 Eddington, *The Nature of the Physical World,* 286.

55 Einstein, "Isaac Newton", 467.

56 Eddington, *The Nature of the Physical World,* 283.

57 Ibid. 295.

58 Ibid. 295–6.

59 The metaphor is not without its problems. For example, it is debatable whether the degree of probability introduced at the quantum level is at all meaningful once at a macro-level, and reducing the ethical, anthropological, sociological, legal and theological arguments regarding human volition and free will to the quantum level is too reductionistic. Yet attempts to work the insights of quantum theory into theology, to appeal to quantum probabilities and uncertainties for a defence of historical and natural novelties, are already well underway. See e.g. Anders S. Tune, "Quantum Theory and the Resurrection of Jesus", *Dialog,* 43/3 (2004): 166–76.

of this shift from strict physical causation opens up a vital and promising space for further investigation and discussion in the science and theology discourse.

Here theology has an advantage over the sciences in that the wealth of its traditions actually offer several conceptions and motifs which reflect a non-deterministic relation between God and creation, for example, discipleship, the kingdom of God, and most importantly the crucified Christ. These metaphors push us beyond thinking of God in absolutist terms, but rather highlight God's openness "to creation, to the world and to man", an openness which provides the framework for "the drama of freedom and love".[60]

Pannenberg's rule *by* God stressed the power and glory of God to direct creation in accordance with his absolute will and by the power of his absolute, all-embracing, and all-determining *potestatem et imperium*[61]—an immense doxology to the illimitable power of God. Yet this picture of God as all-determining reality leads us away from the crucified Christ. In Christ, we do not have the revelation of God as absolute and distant sovereign, whose being is his rule, but as the God who opens up the course of time to human action: to the human desire to be near God and to reject God, to follow God and to turn away. And in our turning away God continues to subject himself to the structure and brokenness of time, not taking refuge outside law and time, to rule by law and judge with time, but to enter in under the law to confront its fallenness rather than *rule* by it. "God is not world providence, ... not a ruler and sovereign of the universe, not *pantokrator*. God is freedom and meaning, love and sacrifice; He is struggle against the objectivized world order. ... He suffers for man and strives together with man against the falsity and wrong of the world, against the intolerable suffering of the world".[62]

Under the law, we are ruled by law, directed and constrained under a deterministic system presided over by an external, determining sovereign. In contrast, under grace we have God's rule of law, a structure which surrounds creation and supports us in freedom, and within which God, as Emmanuel, is also present.

60 Berdyaev, *Slavery and Freedom*, 85.

61 Pannenberg, *Systematic Theology*, i. 417 n. 173.

62 Berdyaev, *Slavery and Freedom*, 89.

If God were solely and purely sovereign then the law would not apply to him.[63] But in the suffering and self-condescension of Christ, God willingly submits himself to it, redeeming it and transforming it.[64] Under an absolutist and imperialist rule *by* God, history is determined towards judgement, and the future holds little hope precisely because it is controlled and instrumentalized by an autocratic and despotic God. Under the rule *of* God, creation receives the supportive framework in which to flourish into the future, and is bestowed God's presence with us together in his kingdom.

Attempts to conceive of God "as Absolute ... [lead to] a monarchical understanding of God ... which is a source of theological seduction and slavery".[65] This road is seductive because it is the path taken by world powers in their clash for absolute dominance, by nations which see ultimate authority solely in terms of absolute power and absolute rule, rejecting all other options as "a strategy of the weak".[66] But "Christianity is not the revelation of God as an absolute monarch":

---

63 "The sovereign is said to be exempt from the law; since, properly speaking, no man is coerced by himself, and law has no coercive power save from the authority of the sovereign. Thus then is the sovereign said to be exempt from the law, because none is competent to pass sentence on him, if he acts against the law"; Thomas Aquinas, *Summa Theologiæ*, 2a, Q96, a5.

64 "But as to the directive force of law, the sovereign [can be] subject to the law by his own will, according to the statement that 'whatever law a man makes for another, he should keep himself'. And a wise authority says: 'Obey the law that thou makest thyself'. Moreover the Lord reproaches those who 'say and do not'; and who 'bind heavy burdens and lay them on men's shoulders, but with a finger of their own they will not move them'. Hence, in the judgement of God, the sovereign is not exempt from the law, as to its directive force; but he should fulfil it to his own free-will"; (ibid.).

65 Berdyaev, *Slavery and Freedom*, 85.

66 "Nevertheless, we have vulnerabilities: ... Our strength as a nation state will continue to be challenged by those who employ *a strategy of the weak* using international fora [*sic*], judicial processes, and terrorism"; United States Department of Defense, *The National Defense Strategy of the United States of America*, March 2005, 5, emphasis added; which publicly equated international diplomatic forums such as the United Nations, and *the rule of law* with acts of terrorism as tools of the "weak" employed against the absolute power of a nation with "no global peer" and "unmatched in traditional military capability" (ibid.).

> The Christian revelation of the Son of God Who sacrificed Himself, suffered and was crucified, saves us from that. God is not an absolute monarch. God is a God Who suffers with the world and with man. He is crucified Love; He is the Liberator. The Liberator appears not as a power but as Crucifixion.[67]

We do not see the perfect revelation of God in the *potentia Dei absoluta* or in God's sovereign reign or deterministic rule over creation via time. We do not see the love of God expressed in the absolute power that dominates everything, in the one who heartlessly becomes the author of all evil and suffering. Seven centuries after Bradwardine, the seemingly infinite truth and boundless challenge of suffering and evil remain the insurmountable hurdles to a shallow and reductionistic theology of God's glory.[68] In the concreteness of a *theologia crucis,* through the suffering and resurrection of Jesus, and in the continuous entry of his Kingdom into the world, we witness the gradual defeat of the fallenness of history and the promise of the crucified Christ's final victory over the brokenness of time.

---

67 Berdyaev, *Slavery and Freedom,* 85.

68 It is on these terms that Moltmann accurately distinguishes between theologians of glory and theologians of the cross: "religious desire for praise and might and self-affirmation are blind to suffering—their own and that of others—because they are in love with success"; Jürgen Moltmann, *The Crucified God* (London: SCM, 1999), 213–14.

BIBLIOGRAPHY

1 Primary Literature—Wolfhart Pannenberg

"Akt und Sein im Mittelalter", in *Kerygma und Dogma*, 7 (1961): 197–220.

"Albertus Magnus: Opera Omnia Vol. XXVIII (ed. Bernhard Geyer), Münster 1951", in *Theologische Literaturzeitung*, 81 (1956): 109–10.

Analogie und Offenbarung. Eine kritische Untersuchung der Geschichte des Analogiebegriffs in der Gotteserkenntnis: Entwurf (typed habilitation thesis: Heidelberg, 1955) = Wolfhart Pannenberg, *Analogie und Offenbarung. Eine kritische Untersuchung zur Geschichte des Analogiebegriffs in der Lehre von der Gotteserkenntnis* (Göttingen: Vandenhoeck & Ruprecht, 2007).

"Analogie", in H. Brunotte and O. Weber (eds.), *Evangelisches Kirchenlexikon* (Göttingen: Vandenhoeck & Ruprecht, 1956), i. 113–14.

"Analogie", in K. Galling et al. (eds.), *Religion in Geschichte und Gegenwart* i. 350–3.

"An Autobiographical Sketch", in C. E. Braaten and P. Clayton (eds.), *The Theology of Wolfhart Pannenberg: Twelve American Critiques with an Autobiographical Essay and Response* (Minneapolis, MN: Augsburg, 1988), 11–18.

"Anselm von Canterbury: Cur Deus Homo. Warum Gott Mensch geworden. Besorgt und übersetzt von F. S. Schmitt, München 1956", in *Theologische Literaturzeitung*, 82 (1957): 203–4.

Anthropology in Theological Perspective, trans. M. J. O'Connell (Edinburgh: T & T Clark, 1999).

"Appearance as the Arrival of the Future", in id., *Theology and the Kingdom of God* (Philadelphia: Westminster Press, 1977), 127–43.

"Atomism, Duration, Form: Difficulties with Process Philosophy", in id., *Metaphysics and the Idea of God* (Grand Rapids: Eerdmans, 1990), 113–29.

Review of "B. Hägglund: Theologie und Philosophie bei Luther und in der ockhamistischen Tradition, Lund 1955", in *Archiv für Reformationsgeschichte*, 47 (1956): 273–5.

Basic Questions in Theology, i (London: SCM, 1970).

"Being and Time", in id., *Metaphysics and the Idea of God* (Grand Rapids, MI: Eerdmans, 1990), 69–90.

Christianity in a Secularized World (London: SCM, 1988).

"Christliche Glaube und menschliche Freiheit", *Kerygma und Dogma*, 4 (1958): 251–80.

"Contingency and Natural Law", in id., *Toward a Theology of Nature: Essays on Science and Faith*, ed. Ted Peters (Louisville, KY: Westminster John Knox, 1993), 72–122.

"Das auserwählte Volk oder: Wird die Nation zum Götzen?", *Deutsches Allgemeines Sonntagsblatt*, 27 (4 July 1976), 26.

"Did Jesus Really Rise From the Dead?", *Dialog*, 4 (1965): 128–35.

"Die Aufgabe einer politischen Theologie des Christentums", in H.-W. Bartsch (ed.), *Kerygma und Mythos VII–1. Glaube und Politik—Religion und Staat. Zur Entmythologisierung und Neubestimmung ihres Verhältnisses* (Hamburg: Reich & Heidrich, 1979), 19–25.

Die Prädestinationslehre des Duns Skotus im Zusammenhang der scholastischen Lehrentwicklung (Forschungen zur Kirchen- und Dogmengeschichte, 4; Göttingen, 1954).

"Dogmatic Theses on the Doctrine of Revelation", in id., *Revelation as History*, trans. D. Granskou (Toronto: Macmillan, 1969), 123–58.

"Eternity, Time, and Space", *Zygon*, 40/1 (March 2005): 97–106.

"God's Presence in History: How My Mind Has Changed", in *Christian Century*, 11 March 1981, 260–3.

Grundzüge der Christologie (Gütersloh: Gütersloher Verlag, 1964) = *Jesus—God and Man*, trans. L. L. Wilkins and D. A. Priebe (Philadelphia: Westminster, 1968).

Review of "H. Lyttkens: The Analogy between God and the World, Uppsala 1952", in Ernst Wolf (ed.), *Verkündigung und Forschung. Theologischer Jahresbericht 1956/57* (München, 1957/59), 136–42.

Review of "H. Mühlen: Sein und Person nach Johannes Duns Scotus, Werl 1954", in *Theologische Literaturzeitung*, 80 (1955): 353–5.

Review of "H. Wagner: Existenz, Analogie und Dialektik, München 1953", in *Theologische Literaturzeitung*, 79 (1954): 318–20.

"Heilsgeschehen und Geschichte I", *Kerygma und Dogma*, 5/3 (1959): 218–37.

"Heilsgeschehen und Geschichte (Fortsetzung und Schluß)", *Kerygma und Dogma*, 5 (1959): 259–88.

"History and the Reality of the Resurrection", in Gavin D'Costa (ed.), *Resurrection Reconsidered* (Oxford: Oneworld, 1996), 62–72.

Human Nature, Election and History (Philadelphia: Westminster, 1977) = Pannenberg, W., *Die Bestimmung des Menschen: Menschsein, Erwählung und Geschichte* (Göttingen: Vandenhoeck & Ruprecht, 1978).

"Introduction", in id. (ed.), *Revelation as History*, trans. D. Granskou (Toronto: Macmillan, 1969), 3–21.

Review of "J. Duns Scotus: Opera Omnia Vol. I–III (ed. Carolus Balic), Civitas Vaticana 1950/54", in *Theologische Literaturzeitung*, 81 (1956): 550–2.

Jesus—God and Man, trans. L. L. Wilkins and D. A. Priebe (2nd edn., Philadelphia: Westminster, 1968).

"Johannes Duns Scotus", in H. Brunotte and O. Weber (eds.), *Evangelisches Kirchenlexikon* (Göttingen: Vandenhoeck & Ruprecht, 1956), 980–2.

"Kerygma und Geschichte", in R. Rendtorff and K. Koch (eds.), *Studien zur Theologie der alttestamentlichen Überlieferungen* (Neukirchen: Neukirchener, 1961), 129–40.

"La signification du Christianisme dans la philosophie de Hegel", *Archives de Philosophie*, 33, 1970 (1971) = "The Significance of Christianity in the Philosophy of Hegel", in id., *The Idea of God and Human Freedom*, trans. R. A. Wilson (Philadelphia: Westminster, 1973), 144–77.

Metaphysics and the Idea of God (Grand Rapids: Eerdmans, 1990).

"Person", in K. Galling et al. (eds.), *Religion in Geschichte und Gegenwart* (Tübingen: JCB Mohr/Siebeck, 1958), v. 230–5.

"Redemptive Event and History", in id., *Basic Questions in Theology*, i (London: SCM, 1970), 15–80.

"Reich Gottes in Amerika. Religiöse Selbstdeutung als Thema politischer Theologie", *Evangelische Kommentare,* 10 (1977): 333–6.

"Reich Gottes und Nationalismus. Vom politishen Sinn der christlichen Hoffnung", in H. J. Schultz (ed.), *Kontexte,* i (Stuttgart: Kreuz, 1965), 41–8.

(ed.), *Revelation as History,* trans. D. Granskou (Toronto: Macmillan, 1969).

Review of "Gordon Leff: Bradwardine and the Pelagians. A Study of his 'De Causa Dei' and its Opponents (= Cambridge Studies in Medieval Life and Thought N. S. vol. 5). Cambridge (University Press) 1957. 282 S.", *Zeitschrift für Kirchengeschichte,* 69 (1958): 355–61.

Review of "Heiko Augustinus Oberman: Archbishop Thomas Bradwardine. A Fourteenth Century Augustinian. A Study of his Theology in its Historical Context (= Diss. Utrecht 1957). Utrecht (Kemink en Zoon) 1957. XI, 246 S., hfl 10.–.", *Zeitschrift für Kirchengeschichte,* 72 (1961): 173–5.

Systematic Theology, trans. G. W. Bromiley, 3 vols. (Grand Rapids, MI: Eerdmans, 1991–97).

"Th. Steinbüchel: Vom Menschenbild des christlichen Mittelalters, Basel 1951", in *Theologische Literaturzeitung,* 81 (1956): 733–5.

"The Biblical Understanding of Reality", in id., *Faith and Reality,* trans. John Maxwell (London: Search, 1977), 8–19.

"The Doctrine of Creation and Modern Science", *Zygon,* 23 (1988): 3–21.

"The Historical Jesus as a Challenge to Christology", *Dialog,* 37/1 (Winter 1998): 22–7.

The Idea of God and Human Freedom, trans. R. A. Wilson (Philadelphia: Westminster, 1973).

"Theological Questions to Scientists", in id., *Toward a Theology of Nature: Essays on Science and Faith,* ed. Ted Peters (Louisville, KY: Westminster John Knox, 1993), 15–28.

"Theologie der Schöpfung und Naturwissenschaft", presentation at the 7th Ökumenischen Sommerakadamie 2005, Linz, 13 July, 2005.

"Theology and the Categories 'Part' and 'Whole'", in id., *Metaphysics and the Idea of God* (Grand Rapids, MI: Eerdmans, 1990), 130–52.

Theology and the Kingdom of God (Philadelphia: Westminster, 1969).

"Theology and the Kingdom of God", in id., *Theology and the Kingdom of God* (Philadelphia: Westminster, 1977), 51–71.

"The Problem of the Absolute", in id., *Metaphysics and the Idea of God* (Grand Rapids, MI: Eerdmans, 1990), 22–42.

Toward a Theology of Nature: Essays on Science and Faith, ed. Ted Peters (Louisville, KY: Westminster John Knox, 1993).

"What is a Dogmatic Statement", in id., *Basic Questions in Theology,* i (London: SCM, 1970), 182–211.

What is Man? Contemporary Anthropology in Theological Perspective, trans. Duane A. Priebe (Philadelphia: Fortress, 1970).

"Zur Bedeutung des Analogiegedankens bei Karl Barth. Eine Auseinandersetzung mit Urs von Balthasar", in *Theologische Literaturzeitung,* 78 (1953): 17–24.

2 Secondary Literature

Augustine, *Confessions and Enchiridion,* trans./ed. A. C. Outler. (Philadelphia: Westminster, 1955).

Austin, J. L., *How to Do Things With Words* (2nd edn.; Cambridge, MA: Harvard University Press, 1975).

Bakhtin, M., *Speech Genres and Other Late Essays,* trans. V. W. McGee (Austin: University of Texas Press, 1986).

Barbour, I. G., *Religion in an Age of Science* (San Francisco: Harper and Row, 1990).

Barbour, I. G., "God's Power: A Process View", in John Polkinghorne (ed.), *The Work of Love: Creation as Kenosis* (Grand Rapids: Eerdmans, 2001), 1–20.

Barth, K., *The Epistle to the Romans,* trans. E. C. Hoskyns (6th edn., London: OUP, 1968).

Bauman, M., *Roundtable: Conversations with European Theologians* (Grand Rapids: Baker, 1990).

Bebbington, D., *Patterns in History: A Christian Perspective on Historical Thought* (Vancouver: Regent College Publishing, 1990).

Beckermann, A., "Introduction—Reductive and Nonreductive Physicalism: A Short Survey of Six Decades of Philosophical Discussion Including an Attempt to Formulate a Version of Physicalism Against the Background of C. D. Broad's Concept of Emergence", in id., Hans Flohr, Jaegwon Kim (eds.), *Emergence or Reduction? Essays on the Prospects of Nonreductive Physicalism* (Berlin: de Gruyter, 1992), 1–21.

Begbie, J. S., *Theology, Music and Time* (Cambridge: Cambridge University Press, 2000).

Berdyaev, N., *Slavery and Freedom*, trans. R. M. French (New York: Charles Scribner's Sons, 1944).

Solitude and Society, trans. G. Reavey (London: Geoffrey Bles, 1947).

The Meaning of History, trans. G. Reavey (London: Bles Centenary, 1936).

Bergson, H., *Time and Free Will: An Essay on the Immediate Data of Consciousness*, trans. F. L. Pogson (Mineola, NY: Dover, 2001).

Bohm, D., "Time, the Implicate Order, and Pre-Space", in D. R. Griffen (ed.), *Physics and the Ultimate Significance of Time: Bohm, Prigogine, and Process Philosophy* (Albany, NY: SUNY Press, 1986), 177–208.

Brown, P., *The Rise of Western Christendom: Triumph and Diversity AD 200–1000* (Cambridge, MA: Blackwell, 1996).

Brown, W. S., N. Murphy, H. N. Malony (eds.), *Whatever Happened to the Soul? Scientific and Theological Portraits of Human Nature* (Minneapolis: Fortress, 1998).

Brown, W. S., "Neurobiological Embodiment of Spirituality and Soul", in Malcolm Jeeves (ed.), *From Cells to Souls—And Beyond: Changing Portraits of Human Nature* (Grand Rapids, MI: Eerdmans, 2004), 58–76.

Brumlik, M., "The Concept of Time and the Faculty of Judgement in the Ontogenesis of Historical Consciousness", in Jürgen Straub (ed.), *Narration, Identity and Historical Consciousness* (New York: Berghahn, 2005), 135–40.

Campenhausen, H. von, "Augustin und der Fall von Rom", in id., *Weltgeschichte und Gottesgericht: zwei akademische Vorträge über Augustine und Luther* (Stuttgart: Kreuz, 1947), 2–18.

Cao, D., *Chinese Law: A Language Perspective* (Aldershot: Ashgate, 2004), 35–55.

Chan, M. L. Y., *Christology from Within and Ahead: Hermeneutics, Contingency, and the Quest for Transcontextual Criteria in Christology* (Leiden: Brill, 2000).

Charles, R. H., "Apocalyptic Literature", in J. Hastings et al. (eds.), *A Dictionary of the Bible Dealing with its Language, Literature and Contents Including the Biblical Theology* (Edinburgh: T & T Clark, 1951).

Chen J., "Market Economy and the Internationalisation of Civil and Commercial Law in the People's Republic of China", in Kanishka Jayasuriya (ed.) *Law, Capitalism and Power in Asia* (London: Routledge, 1999), 69–94.

Cirkovic, M. M., "Is Many Likelier Than Few? A Critical Assessment of the Self-Indication Assumption", *Epistemologia,* 27 (2004): 265–98.

Collingwood, R. G., *An Autobiography* (London: OUP, 1939).

Cotes, R., "The Preface of Mr. Roger Cotes, To the Second Edition of this Work, so far as it relates to the Invention and Discoveries herein contained", in Isaac Newton, *The Mathematical Principles of Natural Philosophy,* trans. Andrew Motte (3rd. edn.; London: Motte, 1729), n.p.

Cowdell, S., *Is Jesus Unique? A Study of Recent Christology* (New York: Paulist, 1996).

Craig, W. L., *Time and Eternity: Exploring God's Relationship to Time* (Wheaton, IL: Crossway, 2001).

Davies, D., *Anthropology and Theology* (Oxford: Berg, 2002).

Davies, P., *About Time: Einstein's Unfinished Revolution* (New York: Touchstone, 1996).

Are We Alone? Philosophical Implications of the Discovery of Extraterrestrial Life (New York: Basic, 1995).

"Teleology without Teleology: Purpose through Emergent Complexity", in Philip Clayton and Arthur Peacocke (eds.), *In Whom We Live And Move And Have Our Being: Panentheistic Reflections on God's Presence in a Scientific World* (Grand Rapids: Eerdmans, 2004), 95–108.

Declaration on Democratic Values, G7 Economic Summit, London, June 8 1984.

Dirac, P. A. M., "Einstein and the Development of Physics", in C. M. Kinnon, A. N. Kholodilin and J. G. Richardson (eds.), *The Impact of Modern Scientific Ideas on Society: In Commemoration of Einstein* (Dordrecht: Reidel, 1981).

Dolnikowski, E. W., *Thomas Bradwardine: A View of Time and a Vision of Eternity in Fourteenth-Century Thought* (Leiden: Brill, 1995).

Douglas, D. C., *English Historical Documents* (Abingdon: Routledge, 2004).

Droysen, J. G., *Historik: Rekonstruktion der ersten vollständigen Fassung der Vorlesungen (1857), Grundriß der Historik in der ersten handschriftlichen (1857/58) und in der letzten gedruckten Fassung (1882)*, ed. by Peter Leyh (Stuttgart: Fromann-Holzboog, 1977).

Dunne, J. W., *An Experiment With Time* (Black, 1927).

Eddington, A., *Space Time and Gravitation: An Outline of the General Relativity Theory* (Cambridge: CUP, 1920).

The Nature of the Physical World (1928; London: Dent, 1964).

Eliade, M., *The Myth of the Eternal Return* (1949; Princeton: PUP, 2005).

The Sacred and the Profane: The Nature of Religion (Orlando, FL: Harcourt, 1987).

Ellis, G. F. R., G. B. Brundrit, "Life in an Infinite Universe", *Quarterly Journal of the Royal Astronomical Society*, 20 (1979): 37–41.

Feynman, R. P., "The Theory of Positrons", *Physical Review*, 76, 749–59 (1949), reprinted in R. P. Feynman, *Quantum Electrodynamics* (Boulder, CO: Westview, 1998).

Fincham, F. D., J. M. Jaspers, "Attribution of Responsibility: From man the scientist to man as lawyer", *Advances in Experimental Social Psychology*, 13 (1980): 81–138.

Flint, R., *History of the Philosophy of History* (New York: Charles Scribner's Sons, 1894).

Freeman, E. A., "The Practical Bearings of European History", in id., *Lectures to American Audiences* (London: Trübner, 1882).

Fuchs, E., "Reshaping the World: Historiography from a Universal Perspective", in Larry E. Jones (ed.), *Crossing Boundaries: The Exclusion and Inclusion of Minorities in Germany and the United States* (New York: Berghahn, 2001), 243–63.

Giedion, S., *Space, Time and Architecture: The Growth of a New Tradition* (Cambridge, MA: Harvard University Press, 1949).

Gilbertson, M., *God and History in the Book of Revelation: New Testament Studies in Dialogue with Pannenberg and Moltmann* (Cambridge: CUP, 2003).

Gilkey, L., *Maker of Heaven and Earth: The Christian Doctrine of Creation in the Light of Modern Knowledge* (Lanham, MD: University Press of America, 1985).

Gilson, É., *Jean Duns Scot. Introduction a ses positions fondamentales* (Paris: Vrin, 1952).

Gloy, K., *Das Verständnis der Natur* (München: C. H. Beck, 1995).

"Die Geschichte des Naturverständnisses", presentation at the 7th Ökumenischen Sommerakadamie 2005, Linz, 14 July 2005.

Grenz, S. J., John R. Franke, *Beyond Foundationalism: Shaping Theology in a Postmodern Context* (Louisville, KY: Westminster John Knox, 2001).

Hacking, I., "The Inverse Gambler's Fallacy: The Argument from Design. The Anthropic Principle Applied to Wheeler Universes", *Mind* NS, 96/383 (Jul., 1987): 331–40.

Hahn, R., "Laplace and the Mechanistic Universe", in David C. Lindberg and Ronald L. Numbers (eds.), *God and Nature: Historical Essays on the Encounter Between Christianity and Science* (Berkeley: University of California Press, 1986), 256–76.

Hamilton, V. L., "Who Is Responsible? Toward a *Social* Psychology of Responsibility Attribution", *Social Psychology,* 41/4 (1978): 316–28.

Haught, J. F., *God After Darwin: A Theology of Evolution* (Boulder, CO: Westview, 2000).

Hawking, S., *A Brief History of Time: From the Big Bang to Black Holes* (London: Bantam, 1988).

Heider, F., *The Psychology of Interpersonal Relations* (New York: Wiley, 1958).

Helm, P., "Eternal Creation: The Doctrine of the Two Standpoints", in C. Gunton (ed.), *The Doctrine of Creation* (Edinburgh: T & T Clark, 1997), 29–46.

Henderson, L. D. "Four-Dimensional Space or Space-Time? The Emergence of the Cubism-Relativity Myth in New York in the 1940s", in Michele Emmer (ed.), *The Visual Mind,* ii (Cambridge, MA: MIT, 2005), 349–98.

Henderson, L. D., *The Fourth Dimension and Non-Euclidean Geometry in Modern Art* (Princeton: PUP, 1983).

Hosinski, T. E., *Stubborn Fact and Creative Advance: An Introduction to the Metaphysics of Alfred North Whitehead* (Lanham, MD: Rowman & Littlefield, 1993).

Hume, D., *An Enquiry Concerning Human Understanding,* ed. T. L. Beauchamp (Oxford: OUP, 1999).

Inwood, M., *A Hegel Dictionary* (The Blackwell Philosopher Dictionaries; Oxford: Blackwell, 1992).

Jammer, M., *Einstein and Religion: Physics and Theology* (Princeton: PUP, 2002).

"Feld—Feldtheorie", in J. Ritter (ed.), *Historisches Wörterbuch der Philosophie* (Basel: Schwabe, 1972), ii. 923–7.

Jeeves, M. (ed.), *From Cells to Souls—And Beyond: Changing Portraits of Human Nature* (Grand Rapids, MI: Eerdmans, 2004).

Jenson, R. W., "Karl Barth", in David Ford (ed.), *The Modern Theologians* (2nd edn., Oxford: Blackwell, 2001), 21–36.

Kandel, E. R., J. H. Schwartz and T. M. Jessell (eds.), *Principles of Neural Science* (4th edn.; New York: McGraw-Hill, 2000).

Kant, I., *Critique of Pure Reason,* trans. J. M. D. Meiklejohn (1781; London and New York: Everyman, 1979).

Kim, J., *Physicalism, Or Something Near Enough* (Princeton: PUP, 2005).

Kirn, O., *Glaube und Geschichte. Eine dogmatische Untersuchung* (Fakultätsprogramm der theologischen Fakultät in Leipzig; Leipzig, 1900).

Koselleck, R., *Vergangene Zukunft: Zur Semantik geschichtlicher Zeiten* (Frankfurt am Main: Suhrkamp, 1989).

Kuhn, T., *The Structure of Scientific Revolution* (Chicago: University of Chicago Press, 1996).

Lakkis, S., "'Have you any right to be angry?' Lament as a metric of socio-political and theological context", in Eva Harasta, Brian Brock (eds.), *Evoking Lament: A Theological Discussion* (Edinburgh: T & T Clark, 2009), 168–82.

"Reclaiming Christian Anthropology in a Post-Soul Era", *Taiwan Journal of Theology,* 37 (2014): in press.

Leff, G., *Bradwardine and the Pelagians: A Study of his "De Causa Dei" and its Opponents* (Cambridge: CUP, 1957).

Lewis, C. S., *The Problem of Pain* (London: Bles Centenary, 1941).

Lindberg, D. C., "Introduction", in id. and Ronald L. Numbers (eds.), *God and Nature: Historical Essays on the Encounter Between Christianity and Science* (Berkeley: University of California Press, 1986), 1–18.

Lorentz, H. A., A. Einstein, and H. Minkowski, *Das Relativitätsprinzip: Eine Sammlung von Abhandlungen mit einem Beitrag von H. Weyl und Anmerkungen von A. Sommerfeld* (Berlin: B.G. Teubner, 1922). = H. A. Lorentz, A. Einstein, H. Minkowski and H. Weyl, *The Principle of Relativity: A Collection of Original Memoirs on the Special and General Theory of Relativity,* trans. W. Perrett and G. B. Jeffrey (London: Methuen, 1923).

Lorenzen, T., *Resurrection and Discipleship: Interpretive Models, Biblical Reflections, Theological Consequences* (Maryknoll, NY: Orbis, 1995).

Luhmann, N., *Die Gesellschaft der Gesellschaft,* ii (Frankfurt am Main: Suhrkamp, 1998).

Lunine, J., "The Frequency of Planetary Systems in the Galaxy", in Ben Zuckerman and Michael Hart (eds.), *Extraterrestrials: Where are They?* (Cambridge: CUP, 1995).

Maclean, I., "Heterodoxy in Natural Philosophy and Medicine: Pietro Pomponazzi, Guglielmo Gratarolo, Girolamo Cardano", in John Brooke, Ian Maclean (eds.), *Heterodoxy in Early Modern Science and Religion* (Oxford: OUP, 2005), 1–30.

Manuel, F. E., *The Religion of Isaac Newton* (Oxford: Clarendon, 1974).

Mattes, M. C., *The Role of Justification in Contemporary Theology* (Grand Rapids: Eerdmans, 2004).

McClendon, Jr., J., *Biography as Theology: How Life Stories Can Remake Today's Theology* (Philadelphia: Trinity, 1990).

McGrath, A. E., *Christian Theology: An Introduction* (Oxford: Blackwell, 2001).

Iustitia Dei: A History of the Christian Doctrine of Justification (3rd edn.; Cambridge: CUP, 2005).

The Intellectual Origins of the European Reformation (Oxford: Blackwell, 2004).

Mildenberger, F., "Kerygmatische Historie?", *Evangelische Theologie* 18 (1958): 419–24.

Minkowski, H., *Raum und Zeit: Vortrag, gehalten auf der 80. Versammlung Deutscher Naturforscher und Ärzte zu Cöln am 21. September 1908* (Leipzig; Berlin: Teubner, 1909); reprinted in H. A. Lorentz, A. Einstein, and H. Minkowski, *Das Relativitätsprinzip: Eine Sammlung von Abhandlungen mit einem Beitrag von H. Weyl und Anmerkungen von A. Sommerfeld* (Berlin: B.G.

Teubner, 1922), 54–66 = H. Minkowski, "Space and Time", in H. A. Lorentz, A. Einstein, H. Minkowski and H. Weyl, *The Principle of Relativity: A Collection of Original Memoirs on the Special and General Theory of Relativity,* trans. W. Perrett and G. B. Jeffrey (London: Methuen, 1923), 73–91.

Moltmann, J., *God in Creation: An Ecological Doctrine of Creation,* trans. M. Kohl (London: SCM, 1997).

The Crucified God, trans. R. A. Wilson, J. Bowden, M. Kohl (London: SCM, 1999), 213–14.

Theology of Hope: On the Ground and the Implications of a Christian Eschatology, trans. J. W. Leitch (London: SCM, 2002).

Mutschler, H.-D., "Schöpfungstheologie und physikalischer Feldbegriff bei Wolfhart Pannenberg", *Theologie und Philosophie,* 70 (1995): 543–58.

Müller, A. M. K., *Das unbekannte Land: Konflikt-Fall Natur; Erfahrungen und Visionen im Horizont der offenen Zeit* (Stuttgart: Radius, 1987).

Nanda, M. "The Epistemic Charity of the Social Constructivist Critics of Science and Why the Third World Should Refuse the Offer", in Noretta Koertge (ed.), *A House Built on Sand: Exposing Postmodernist Myths About Science* (Oxford: Oxford University Press, 1998), 286–308.

Neie, H., *The Doctrine of the Atonement in the Theology of Wolfhart Pannenberg* (Berlin: Walter de Gruyter, 1978).

Newton, I., *The Mathematical Principles of Natural Philosophy,* trans. Andrew Motte (3rd. edn.; London: Motte, 1729).

Oberman, H. A., *Archbishop Thomas Bradwardine. A Fourteenth Century Augustinian. A Study of his Theology in its Historical Context* (Utrecht: Kemink en Zoon, 1957).

O'Regan, C., *The Heterodox Hegel* (Albany: SUNY Press, 1994).

Palmer, R. E., *Hermeneutics: Interpretation Theory in Schleiermacher, Dilthey, Heidegger, and Gadamer* (Evanston, IL: Northwestern University Press, 1969).

Peerenboom, R., "Competing Conceptions of Rule of Law in China", in id. (ed.), *Asian Discourses of Rule of Law: Theories and Implementation of Rule of Law in Twelve Asian Countries, France and the U.S.* (London: Routledge, 2004), 109–41.

Peters, T., "Naturwissenschaft und Religion. Ein wachsender Forschungsbereich", *Verkündigung und Forschung,* 49 (2/2004): 52–72.

"Pannenberg on Theology and Natural Science", in Wolfhart Pannenberg, *Toward a Theology of Nature: Essays on Science and Faith,* ed. Ted Peters (Louisville, KY: Westminster/John Knox Press, 1993), 1–14.

"Science and Theology: Toward Consonance", in id. (ed.), *Science and Theology: The New Consonance* (Boulder, CO: Westview, 1998), 11–39.

Picht, G., "Die Zeit und die Modalitäten", in H. P. Dürr (ed.), *Quanten und Felder. Physikalische und philosophische Betrachtungen zum 70. Geburtstag Werner Heisenberg* (Brunswick: Vieweg, 1971), 67–76.

Placher, W., *A History of Christian Theology* (Philadelphia: Westminster, 1983).

Polkinghorne, J., *Belief in God in an Age of Science* (New Haven: Yale Nota Bene, 1998).

with Michael Welker (eds.), *The End of the World and the Ends of God: Science and Theology on Eschatology* (Harrisburg, PA: Trinity, 2000).

with Michael Welker, "Introduction: Science and Theology on the End of the World and the Ends of God", in eid. (eds.), *The End of the World and the Ends of God: Science and Theology on Eschatology* (Harrisburg, PA: Trinity, 2000), 1–16.

Reason and Reality: The Relationship between Science and Theology (Philadelphia: Trinity, 1991).

Preuss, H. D., *Old Testament Theology,* ii, trans. Leo G. Perdue (Louisville, KY: Westminster John Knox, 1995).

Rad, G. von, *Theologie des Alten Testaments,* i (4th edn., Munich: Kaiser, 1965).

Robinson, J. M., "Revelation as Word and as History", in id. and John B. Cobb, Jr. (eds.), *Theology as History* (New Frontiers in Theology: Discussions among Continental and American Theologians, 3; New York: Harper & Row, 1967), 1–100.

Rüsch, T., *Die Entstehung der Lehre vom Heiligen Geist bei Ignatius von Antiochia, Theophilus von Antiochia and Irenäus von Lyon* (Zürich: Zwingli, 1952).

Sanders, J., and V. Lee Hamilton (eds.), *Handbook of Justice Research in Law* (New York: Kluwer Academic/Plenum, 2001).

Schleiermacher, F. D. E., *The Christian Faith*, ed. H. R. Mackintosh and J. S. Stewart, trans. D. M. Baillie et al. (Edinburgh: T & T Clark, 1928).

Schwöbel, C., "The Church as a Cultural Space: Eschatology and Ecclesiology", in John Polkinghorne and Michael Welker (eds.), *The End of the World and the Ends of God: Science and Theology on Eschatology* (Harrisburg, PA: Trinity, 2000), 107–23.

Schüle, A. (ed.), *Biography as Religious and Cultural Text* (Münster: LIT, 2002).

Seiwert, H., "End of Time and New Time in Medieval Chinese Buddhism", in A. I. Baumgarten (ed.), *Apocalyptic Time* (Leiden: Brill, 2000), 1–14.

Shults, F. L., *The Postfoundationalist Task of Theology: Wolfhart Pannenberg and the New Theological Rationality* (Grand Rapids: Eerdmans, 1999).

Smart, J. J. C., "Time", in Paul Edwards (ed.), *Encyclopaedia of Philosophy* (London: Collier-Macmillan, 1967), viii. 126.

Spengler, O., *Decline of the West*, trans. C. F. Atkinson (Oxford: Oxford University Press, 1991).

Spinoza, B., "Tractatus theologico-politicus", in id., *The Chief Works of Benedict de Spinoza*, trans. R. H. M. Elwes (London: George Bell, 1909).

Stannard, R., "God in and beyond Space and Time", in Philip Clayton and Arthur Peacocke (eds.), *In Whom We Live And Move And Have Our Being: Panentheistic Reflections on God's Presence in a Scientific World* (Grand Rapids: Eerdmans, 2004), 109–20.

Stiver, D. R., *The Philosophy of Religious Language: Sign Symbol and Story* (Oxford: Blackwell, 2002).

Straub, J., "Telling Stories, Making History: Toward a Narrative Psychology of the Historical Construction of Meaning", in id. (ed.), *Narration, Identity and Historical Consciousness* (New York: Berghahn, 2005), 44–98.

Strong, J. S., *Relics of the Buddha* (Princeton: PUP, 2004).

Tamanaha, B. Z., "The Rule of Law for Everyone?", *Current Legal Problems*, 55 (2002): 97–122; also available at <http://ssrn.com/abstract=312622>.

Thomas Aquinas, *Summa Theologiæ* (Blackfriars edn.; London: Eyre & Spottiswoode, 1964–81).

Thompson, E. P., *Whigs and Hunters: The Origin of the Black Act* (London: Allen Lane, 1975).

Toit, C. du, "Fin de siécle and the development of a post-secular religion: intimations from the TRC", *Religion and Theology*, 4/1 (1997), available online from the University of South Africa <http://www.unisa.ac.za/default.asp?Cmd=ViewContent &ContentID=7368>, accessed 20 June 2007.

Torrance, A. J., "Creatio Ex Nihilo and the Spatio-Temporal Dimensions, with special reference to Jürgen Moltmann and D. C. Williams", in Colin E. Gunton (ed.), *The Doctrine of Creation: Essays in Dogmatics, History and Philosophy* (Edinburgh: T & T Clark, 1997), 83–103.

Torrance, T. F., *Space, Time and Incarnation* (Oxford: OUP, 1969).

Troeltsch, E., "Ueber Historische und dogmatische Methode in der Theologie", in id., *Gesammelte Schriften*, ii (Tübingen: J. C. B. Mohr [Paul Siebeck], 1913), 729–753.

Trost, L. A., "Non-interventionist Divine Action: Robert Russell, Wolfhart Pannenberg, and the Freedom of the (Natural) World", in Ted Peters and Nathan Hallanger (eds.), *God's Action in Nature's World: Essays in Honour of Robert John Russell* (Aldershot: Ashgate, 2006), 205–216.

Tune, A. S., "Quantum Theory and the Resurrection of Jesus", *Dialog*, 43/3 (2004): 166–76.

Tyndall, J., *Light and Electricity: Notes of Two Courses of Lectures Before the Royal Institution of Great Britain* (New York: D. Appleton, 1871).

United States Department of Defense, *The National Defense Strategy of the United States of America*, March 2005.

Vallée, G., *The Shaping of Christianity: The History and Literature of its Formative Centuries (100–800)* (New York: Paulist Press, 1999).

Vanstone, W. H., *Love's Endeavour, Love's Expense* (London: Darton, Longman and Todd, 1977).

Vignaux, P., *Nominalisme au XIVe siècle* (Paris: Vrin, 1948).

Walker, J. T., *The Legend of Mar Qardagh: Narrative and Christian Heroism in Late Antique Iraq* (Berkeley: University of California Press, 2006).

Watts, M. J., "Capitalisms, Crises, and Cultures 1: Notes Toward a Totality of Fragments", in Allan Pred and Michael J. Watts, *Reworking Modernity: Capitalisms and Symbolic Discontent* (New Brunswick, NJ: Rutgers University Press, 1992), 1–20.

Weinert, F., *The Scientist as Philosopher: Philosophical Consequences of Great Scientific Discoveries* (Berlin: Springer, 2005).

Welker, M., "Introduction", in id. (ed.), *The Work of The Spirit: Pneumatology and Pentecostalism* (Grand Rapids, MI: Eerdmans, 2006), pp. ix–xviii.

Whitaker, A., *Einstein, Bohr and the Quantum Dilemma: From Quantum Theory to Quantum Information* (2nd edn.; Cambridge: CUP, 1996).

Whitehead, A. N., *Process and Reality: An Essay in Cosmology* (corrected edn.; New York: Free Press, 1978).

Science and the Modern World (New York: Free Press, 1967).

The Principle of Relativity with Applications to Physical Science (Cambridge: CUP, 1922).

Wicken, J. S., "Theology and Science in the Evolving Cosmos: A Need for Dialogue", *Zygon*, 23 (1988): 45–55.

Williams, D. C., "The Myth of Passage", *The Journal of Philosophy*, 48/15 (1951): 457–72.

Wilson, E. O., *Sociobiology: The New Synthesis* (Cambridge, MA: Belknap, 1976).

Zeilinger, A., "Encounters Between Buddhist and Quantum Epistemologies", in B. Alan Wallace (ed.), *Buddhism and Science: Breaking New Ground* (New York: Columbia University Press, 2003), 387–97.

INDEX OF NAMES